CAMBRIDGE LIBRARY COLLECTION
Books of enduring scholarly value

History

The books reissued in this series include accounts of historical events and movements by eye-witnesses and contemporaries, as well as landmark studies that assembled significant source materials or developed new historiographical methods. The series includes work in social, political and military history on a wide range of periods and regions, giving modern scholars ready access to influential publications of the past.

The Travels and Researches of Alexander von Humboldt

In 1832, William MacGillivray published this abridged version of the explorer and naturalist Alexander von Humboldt's Personal Narrative of Travels to the Equinoctial Regions of the New Continent During the Years 1799-1804, which had appeared in a seven-volume English translation between 1814 and 1829. MacGillivray's edition, intended for the general public, also includes Humboldt's accounts of his explorations of the Ural Mountains and Caspian Sea. Humboldt became a major figure in physical geography as a result of his arduous five-year trip to explore Central and South America. This book offers a brief biographical sketch of the scientist and covers his exciting journeys from the Island of Tenerife across the Atlantic Ocean to Caracas, and up the Orinoco River by canoe. Humboldt fights mosquitoes in dense rain forests and climbs Andean peaks in Peru without mountain gear, taking detailed notes at every stage.

Cambridge University Press has long been a pioneer in the reissuing of out-of-print titles from its own backlist, producing digital reprints of books that are still sought after by scholars and students but could not be reprinted economically using traditional technology. The Cambridge Library Collection extends this activity to a wider range of books which are still of importance to researchers and professionals, either for the source material they contain, or as landmarks in the history of their academic discipline.

Drawing from the world-renowned collections in the Cambridge University Library, and guided by the advice of experts in each subject area, Cambridge University Press is using state-of-the-art scanning machines in its own Printing House to capture the content of each book selected for inclusion. The files are processed to give a consistently clear, crisp image, and the books finished to the high quality standard for which the Press is recognised around the world. The latest print-on-demand technology ensures that the books will remain available indefinitely, and that orders for single or multiple copies can quickly be supplied.

The Cambridge Library Collection will bring back to life books of enduring scholarly value (including out-of-copyright works originally issued by other publishers) across a wide range of disciplines in the humanities and social sciences and in science and technology.

The Travels and Researches of Alexander von Humboldt

Being a Condensed Narrative of his Journeys in the Equinoctial Regions of America, and in Asiatic Russia; Together with Analyses of his More Important Investigations

WILLIAM MACGILLIVRAY

CAMBRIDGE UNIVERSITY PRESS

Cambridge, New York, Melbourne, Madrid, Cape Town, Singapore,
São Paolo, Delhi, Dubai, Tokyo

Published in the United States of America by Cambridge University Press, New York

www.cambridge.org
Information on this title: www.cambridge.org/9781108004626

© in this compilation Cambridge University Press 2009

This edition first published 1832
This digitally printed version 2009

ISBN 978-1-108-00462-6 Paperback

This book reproduces the text of the original edition. The content and language reflect the beliefs, practices and terminology of their time, and have not been updated.

Cambridge University Press wishes to make clear that the book, unless originally published by Cambridge, is not being republished by, in association or collaboration with, or with the endorsement or approval of, the original publisher or its successors in title.

Engraved by J. Horsburgh

BARON F. H. A. HUMBOLDT.

PUBLISHED BY OLIVER & BOYD, EDINBURGH.

TRAVELS AND RESEARCHES

OF

BARON HUMBOLDT.

OLIVER & BOYD, EDINBURGH.

THE

TRAVELS AND RESEARCHES

OF

ALEXANDER VON HUMBOLDT:

BEING A CONDENSED

NARRATIVE OF HIS JOURNEYS

IN THE

EQUINOCTIAL REGIONS OF AMERICA,

AND IN

ASIATIC RUSSIA;

TOGETHER WITH

ANALYSES OF HIS MORE IMPORTANT
INVESTIGATIONS.

BY W. MACGILLIVRAY, A. M.,

Conservator of the Museums of the Royal College of Surgeons of Edinburgh, Member of
the Natural History Societies of Edinburgh and Philadelphia, &c.

WITH A PORTRAIT OF HUMBOLDT BY HORSBURGH, A MAP OF THE ORINOCO
BY BRUCE, AND FIVE ENGRAVINGS BY JACKSON.

EDINBURGH:
OLIVER & BOYD, TWEEDDALE COURT;
AND SIMPKIN & MARSHALL, LONDON.

MDCCCXXXII.

PREFACE.

The celebrity which Baron Humboldt enjoys, and which he has earned by a life of laborious investigation and perilous enterprise, renders his name familiar to every person whose attention has been drawn to political statistics or natural philosophy. In the estimation of the learned no author of the present day occupies a higher place among those who have enlarged the boundaries of human knowledge. To every one accordingly whose aim is the general cultivation of the mental faculties, his works are recommended by the splendid pictures of scenery which they contain, the diversified information which they afford respecting objects of universal interest, and the graceful attractions with which he has succeeded in investing the majesty of science.

These considerations have induced the Publishers to offer a condensed account of his Travels and Researches, such as, without excluding subjects even of laboured investigation, might yet chiefly embrace those which are best suited to the purposes of the general reader. The public taste has of late years gradually inclined towards objects of useful knowledge,—works of imagination have in a great measure given place to those occupied with descriptions of nature, physical or moral,—and the phenomena of the material world now afford entertainment to many who in former times would have sought for it at a different source. Romantic incidents, perilous adventures, the struggles of conflicting armies, and vivid delineations of national manners and individual character, naturally excite a lively interest in every bosom, whatever may be the age or sex; but, surely, the great facts of creative power and wisdom, as exhibited in regions of the globe of which they have no personal knowledge, are not less calculated to fix the attention of all reflecting minds. The magnificent vegetation of the tropical regions, displaying forests of gigantic trees, interspersed with the varied foliage of innumerable shrubs, and adorned with festoons of climbing and odoriferous plants; the elevated table-lands of the Andes, crowned by volcanic cones, whose summits shoot high into the region of perennial snow; the earthquakes that have desolated populous and fertile countries; the vast expanse of the Atlantic Ocean, with its circling currents; and the varied aspect of the heavens in those distant lands,—are subjects suited to the taste of every individual who is capable of contemplating the wonderful machinery of the universe.

It is unnecessary here to present an analysis of the labours of the illustrious philosopher whose footsteps are traced in this volume. Suffice it to observe, that some notices respecting his early life introduce the reader to an acquaintance with his character and motives, as the adventurous traveller, who, crossing the Atlantic, traversed the ridges and plains of Venezuela, ascended the Orinoco to its junction with the Amazon, sailed down the former river to the capital of Guiana, and after examining the Island of Cuba mounted by the valley of the Magdalena to the elevated platforms of the Andes, explored the majestic solitudes of the great cordilleras of Quito, navigated the margin of the Pacific Ocean, and wandered over the extensive and interesting provinces of New Spain, whence he made his way back by the United States to Europe. The publication of the important results of this journey was not completed when he undertook another to Asiatic Russia and the confines of China, from which he has but lately returned.

From the various works which he has given to the world have been derived the chief materials of this narrative; and, when additional particulars were wanted, application was made to M. de Humboldt himself, who kindly pointed out the sources whence the desired information might be obtained. The life of a man of letters, he justly observed, ought to be sought for in his books; and for this reason little has been said respecting his occupations during the intervals of repose which have succeeded his perilous journeys.

It is only necessary further to apprize the reader, that the several measurements, the indications of the thermometer, and the value of articles of industry or commerce, which in the original volumes are expressed according to French, Spanish, and Russian usage, have been reduced to English equivalents.

Finally, the Publishers, confident that this abridged account of the travels of Humboldt will prove beneficial in diffusing a knowledge of the researches of that eminent naturalist, and in leading to the study of those phenomena which present themselves daily to the eye, send it forth with a hope that its reception will be as favourable and extensive as that bestowed upon its predecessors.

EDINBURGH, *October* 1832.

CONTENTS.

CHAPTER I.
INTRODUCTION.

Birth and Education of Humboldt—His early Occupations—He resolves to visit Africa—Is disappointed in his Views, and goes to Madrid, where he is introduced to the King and obtains Permission to visit the Spanish Colonies—Observations made on the Journey through Spain—Geological Constitution of the Country between Madrid and Corunna—Climate—Ancient Submersion of the Shores of the Mediterranean—Reception at Corunna, and Preparations for the Voyage to South America,............Page 17

CHAPTER II.
VOYAGE FROM CORUNNA TO TENERIFFE.

Departure from Corunna—Currents of the Atlantic Ocean—Marine Animals—Falling Stars—Swallows—Canary Islands—Lancerota—Fucus vitifolius—Causes of the Green Colour of Plants—La Graciosa—Stratified Basalt alternating with Marl—Hyalite—Quartz Sand—Remarks on the Distance at which Mountains are visible at Sea, and the Causes by which it is modified—Landing at Teneriffe,... 25

CHAPTER III.
ISLAND OF TENERIFFE.

Santa Cruz—Villa de la Laguna—Guanches—Present Inhabitants of Teneriffe—Climate—Scenery of the Coast—Orotava—Dragon-tree—Ascent of the Peak—Its Geological Character—Eruptions—Zones of Vegetation—Fires of St John,....................... 41

CHAPTER IV.

PASSAGE FROM TENERIFFE TO CUMANA.

Departure from Santa Cruz—Floating Seaweeds—Flying-fish—Stars—Malignant Fever—Island of Tobago—Death of a Passenger—Island of Coche—Port of Cumana—Observations made during the Voyage; Temperature of the Air; Temperature of the Sea; Hygrometrical State of the Air; Colour of the Sky and Ocean,..Page 55

CHAPTER V.

CUMANA.

Landing at Cumana—Introduction to the Governor—State of the Sick—Description of the Country and City of Cumana—Mode of Bathing in the Manzanares—Port of Cumana—Earthquakes; Their Periodicity; Connexion with the State of the Atmosphere; Gaseous Emanations; Subterranean Noises; Propagation of Shocks; Connexion between those of Cumana and the West Indies; and General Phenomena,.. 68

CHAPTER VI.

RESIDENCE AT CUMANA.

Lunar Halo—African Slaves—Excursion to the Peninsula of Araya—Geological Constitution of the Country—Salt-works of Araya—Indians and Mulattoes—Pearl-fishery—Maniquarez—Mexican Deer—Spring of Naphtha,.. 77

CHAPTER VII.

MISSIONS OF THE CHAYMAS.

Excursion to the Missions of the Chayma Indians—Remarks on Cultivation—The Impossible—Aspect of the Vegetation—San

Fernando—Account of a Man who suckled a Child—Cumanacoa —Cultivation of Tobacco—Igneous Exhalations—Jaguars— Mountain of Cocollar—Turimiquiri—Missions of San Antonio and Guanaguana,..Page 85

CHAPTER VIII.

EXCURSION CONTINUED, AND RETURN TO CUMANA.

Convent of Caripe—Cave of Guacharo, inhabited by Nocturnal Birds—Purgatory—Forest Scenery—Howling Monkeys—Vera Cruz—Cariaco — Intermittent Fevers — Cocoa-trees — Passage across the Gulf of Cariaco to Cumana,............................ 99

CHAPTER IX.

INDIANS OF NEW ANDALUSIA.

Physical Constitution and Manners of the Chaymas—Their Languages—American Races,..111

CHAPTER X.

RESIDENCE AT CUMANA.

Residence at Cumana—Attack of a Zambo—Eclipse of the Sun— Extraordinary Atmospherical Phenomena—Shocks of an Earthquake—Luminous Meteors,..121

CHAPTER XI.

VOYAGE FROM CUMANA TO GUAYRA.

Passage from Cumana to La Guayra—Phosphorescence of the Sea— Group of the Caraccas and Chimanas—Port of New Barcelona— La Guayra—Yellow Fever—Coast and Cape Blanco—Road from La Guayra to Caraccas,...128

CHAPTER XII.

CITY OF CARACCAS AND SURROUNDING DISTRICT.

City of Caraccas—General View of Venezuela—Population—Climate—Character of the Inhabitants of Caraccas—Ascent of the Silla—Geological Nature of the District and the Mines, Page 143

CHAPTER XIII.

EARTHQUAKES OF CARACCAS.

Extensive Connexion of Earthquakes—Eruption of the Volcano of St Vincent's—Earthquake of the 26th March 1812—Destruction of the City—Ten Thousand of the Inhabitants killed—Consternation of the Survivors—Extent of the Commotions,......157

CHAPTER XIV.

JOURNEY FROM CARACCAS TO THE LAKE OF VALENCIA.

Departure from Caraccas—La Buenavista—Valleys of San Pedro and the Tuy—Manterola—Zamang-tree—Valleys of Aragua—Lake of Valencia—Diminution of its Waters—Hot Springs—Jaguar—New Valencia—Thermal Waters of La Trinchera—Porto Cabello—Cow-tree—Cocoa-plantations—General View of the Littoral District of Venezuela,.................................166

CHAPTER XV.

JOURNEY ACROSS THE LLANOS, FROM ARAGUA TO SAN FERNANDO.

Mountains between the Valleys of Aragua and the Llanos—Their Geological Constitution—The Llanos of Caraccas—Route over the Savannah to the Rio Apure—Cattle and Deer—Vegetation—Calabozo—Gymnoti or Electric Eels—Indian Girl—Alligators and Boas—Arrival at San Fernando de Apure,..................186

CHAPTER XVI.

VOYAGE DOWN THE RIO APURE.

San Fernando—Commencement of the Rainy Season—Progress of Atmospherical Phenomena—Cetaceous Animals—Voyage down the Rio Apure—Vegetation and Wild Animals—Crocodiles, Chiguires, and Jaguars—Don Ignacio and Donna Isabella—Water-fowl—Nocturnal Howlings in the Forest—Caribe-fish—Adventure with a Jaguar—Manatees—Mouth of the Rio Apure,..Page 202

CHAPTER XVII.

VOYAGE UP THE ORINOCO.

Ascent of the Orinoco—Port of Encaramada—Traditions of a Universal Deluge—Gathering of Turtles' Eggs—Two Species described—Mode of collecting the Eggs and of manufacturing the Oil—Probable Number of these Animals on the Orinoco—Decorations of the Indians—Encampment of Pararuma—Height of the Inundations of the Orinoco—Rapids of Tabage,..................219

CHAPTER XVIII.

VOYAGE UP THE ORINOCO CONTINUED.

Mission of Atures—Epidemic Fevers—Black Crust of Granitic Rocks—Causes of Depopulation of the Missions—Falls of Apures—Scenery—Anecdote of a Jaguar—Domestic Animals—Wild Man of the Woods—Mosquitoes and other poisonous Insects—Mission and Cataracts of Maypures—Scenery—Inhabitants—Spice-trees—San Fernando de Atabipo—San Baltasar—The Mother's Rock—Vegetation—Dolphins—San Antonio de Javita—Indians—Elastic Gum—Serpents—Portage of the Pimichin—Arrival at the Rio Negro, a Branch of the Amazon—Ascent of the Casiquiare,..239

CHAPTER XIX.

ROUTE FROM ESMERALDA TO ANGOSTURA.

Mission of Esmeralda—Curare Poison—Indians—Duida Mountain—Descent of the Orinoco—Cave of Ataruipe—Raudalito of Carucari—Mission of Uruana—Character of the Otomacs—Clay eaten by the Natives—Arrival at Angostura—The Travellers attacked by Fever—Ferocity of the Crocodiles,....Page 272

CHAPTER XX.

JOURNEY ACROSS THE LLANOS TO NEW BARCELONA.

Departure from Angostura—Village of Cari—Natives—New Barcelona—Hot Springs—Crocodiles—Passage to Cumana, ...288

CHAPTER XXI.

PASSAGE TO HAVANNAH, AND RESIDENCE IN CUBA.

Passage from New Barcelona to Havannah—Description of the latter—Extent of Cuba—Geological Constitution—Vegetation—Climate—Population—Agriculture—Exports—Preparations for joining Captain Baudin's Expedition—Journey to Batabano, and Voyage to Trinidad de Cuba,......................................298

CHAPTER XXII.

VOYAGE FROM CUBA TO CARTHAGENA.

Passage from Trinidad of Cuba to Carthagena—Description of the latter—Village of Turbaco—Air-volcanoes—Preparations for ascending the Rio Magdalena,......................................309

CHAPTER XXIII.

BRIEF ACCOUNT OF THE JOURNEY FROM CARTHAGENA TO QUITO AND MEXICO.

Ascent of the Rio Magdalena—Santa Fe de Bogota—Cataract of Tequendama—Natural Bridges of Icononzo—Passage of Quindiu—Cargueros—Popayan—Quito—Cotopaxi and Chimborazo—Route from Quito to Lima—Guayaquil—Mexico—Guanaxuato—Volcano of Jorullo—Pyramid of Cholula,..................Page 323

CHAPTER XXIV.

DESCRIPTION OF NEW SPAIN OR MEXICO.

General Description of New Spain or Mexico—Cordilleras—Climates—Mines—Rivers—Lakes—Soil—Volcanoes—Harbours—Population—Provinces—Valley of Mexico, and Description of the Capital—Inundations, and Works undertaken for the Purpose of preventing them,..................343

CHAPTER XXV.

STATISTICAL ACCOUNT OF NEW SPAIN CONTINUED.

Agriculture of Mexico—Banana, Manioc, and Maize—Cereal Plants—Nutritive Roots and Vegetables—Agave Americana—Colonial Commodities—Cattle and Animal Productions,.......375

CHAPTER XXVI.

MINES OF NEW SPAIN.

Mining Districts—Metalliferous Veins and Beds—Geological Relations of the Ores—Produce of the Mines—Recapitulation,..390

CHAPTER XXVII.

PASSAGE FROM VERA CRUZ TO CUBA AND PHILADELPHIA, AND VOYAGE TO EUROPE.

Departure from Mexico—Passage to Havannah and Philadelphia—Return to Europe—Results of the Journeys in America, Page 401

CHAPTER XXVIII.

JOURNEY TO ASIA.

Brief Account of Humboldt's Journey to Asia, with a Sketch of the Four great Chains of Mountains which intersect the Central Part of that Continent,...407

ENGRAVINGS.

PORTRAIT OF BARON F. H. A. HUMBOLDT,—*To face the Vignette.*
VIGNETTE—Basaltic Rocks and Cascade of Regla.
Dragon-tree of Orotava,...*Page* 48
Humboldt's Route on the Orinoco,...................................129
Jaguar, or American Tiger,...212
Air-volcanoes of Turbaco,..318
Costumes of the Indians of Mechoacan,..............................341

THE
TRAVELS AND RESEARCHES
OF
BARON HUMBOLDT.

CHAPTER I.

Introduction.

Birth and Education of Humboldt—His early Occupations—He resolves to visit Africa—Is disappointed in his Views, and goes to Madrid, where he is introduced to the King, and obtains Permission to visit the Spanish Colonies—Observations made on the Journey through Spain—Geological Constitution of the Country between Madrid and Corunna—Climate—Ancient Submersion of the Shores of the Mediterranean—Reception at Corunna, and Preparations for the Voyage to South America.

WITH the name of Humboldt we associate all that is interesting in the physical sciences. No traveller who has visited remote regions of the globe, for the purpose of observing the varied phenomena of nature, has added so much to our stock of positive knowledge. While the navigator has explored the coasts of unknown lands, discovered islands and shores, marked the depths of the sea, estimated the force of currents, and noted the more obvious traits in the aspect of the countries at which he has touched; while the zoologist has investigated the multiplied forms of animal life, the botanist the diversified vegetation, the geologist the structure and relations of the rocky masses of which the exterior of the earth is composed; and while each has thus contributed to the illustration of the wonderful constitution of our planet, the distinguished traveller whose discoveries form the subject of this volume stands alone as uniting in himself a knowledge of all these sciences. Geography, meteorology, magnetism, the distribution of heat, the various departments of natural history, together with the affinities of races and languages, the history of nations, the political constitution of countries, statistics, commerce, and agriculture,—all have received accumulated and valuable additions from the exercise of his rare talents. The narrative of no traveller therefore could be more interesting to the man of varied information. But as from a work like that of which the present volume constitutes a part subjects strictly scientific must be excluded, unless when they can be treated in a manner intelligible to the public at large, it may here be stated, that many of the investigations of which we present the results, must be traced in the voluminous works which the author himself has published. At the same time enough will be given to gratify the scientific reader; and while the narrative of personal adventure, the diversified phenomena of the physical world, the condition of societies, and the numerous other subjects discussed, will afford amusement and instruction, let it be remembered that truths faithfully extracted from the book of nature are alone calculated to enlarge the sphere of mental vision; and that, while fanciful description is more apt to mislead than to direct the footsteps of the student, there is reflected from the actual examination of the material universe a light which never fails to conduct the mind at once to sure knowledge and to pious sentiment.

Frederick Henry Alexander Von Humboldt was born at Berlin on the 14th of September 1769. He received his academic education at Göttingen and Frankfort on the Oder. In 1790 he visited Holland and England in company with Messrs George Forster and Van Geuns, and in the same year published his first work, entitled " Observations on the Basalts of the Rhine." In 1791 he went to Freyberg to receive the instructions of the celebrated Werner, the founder of geological science. The results of some of his observations in the mines of that district were published in 1793, under the title of *Specimen Floræ Fribergensis Subterraneæ*.

Having been appointed assessor of the Council of Mines at Berlin in 1792, and afterwards director-general of the mines of the principalities of Baireuth and Anspach in Franconia, he directed his efforts to the formation of public establishments in these districts; but in 1795 he resigned his office with the view of travelling, and visited part of Italy. His active and comprehensive mind engaged in the study of all the physical sciences; but the discoveries of Galvani seem at this period to have more particularly attracted his attention. The results of his experiments on animal electricity were published in 1796, with notes by Professor Blumenbach. In 1795 he had gone to Vienna, where he remained some time, ardently engaged in the study of a fine collection of exotic plants in that city. He travelled through several cantons of Salzburg and Styria with the celebrated Von Buch, but was prevented by the war which then raged in Italy from extending his journey to that country, whither he was anxious to proceed for the purpose of examining the volcanic districts of Naples and Sicily. Accompanied by his brother William Von Humboldt and Mr Fischer, he then visited Paris, where he formed an acquaintance with M. Aimé Bonpland, a pupil of the School of Medicine and Garden of Plants, who, afterwards becoming his associate in travel, has greatly distinguished himself by his numerous discoveries in botany.

Humboldt, from his earliest youth, had cherished an ardent desire to travel into distant regions little known to Europeans, and, having at the age of eighteen resolved to visit the New Continent, he prepared himself by examining some of the most interesting parts of Europe, that he might be enabled to compare the geological structure of these two portions of the globe, and acquire a practical acquaintance with the instruments best adapted for aiding him in his observations. Fortunate in possessing ample pecuniary resources, he did not experience the privations which have disconcerted the plans and retarded the progress of many eminent individuals; but, not the less subject to unforeseen vicissitudes, he had to undergo several disappointments that thwarted the schemes which, like all men of ardent mind, he had indulged himself in forming. Meeting with a person passionately fond of the fine arts, and anxious to visit Upper Egypt, he resolved to accompany him to that interesting country; but political events interfered and forced him to abandon the project. The knowledge of the monuments of the more ancient nations of the Old World, which he acquired at this period, was subsequently of great use to him in his researches in the New Continent. An expedition of discovery to the southern hemisphere, under the direction of Captain Baudin, then preparing in France, and with which MM. Michaux and Bonpland were to be associated as naturalists, held out to him the hope of gratifying his desire of exploring unknown regions. But the war which broke out in Germany and Italy compelled the government to withdraw the funds allotted to this enterprise. Becoming acquainted with a Swedish consul who happened to pass through Paris, with the view of embarking at Marseilles on a mission to Algiers, he resolved to embrace the opportunity thus offered of visiting Africa, in order to examine the lofty chain of mountains in the empire of Morocco, and ultimately to join the body of scientific men attached to the French army in Egypt. Accompanied by his friend Bonpland, he therefore betook himself to Marseilles, where he waited for two months the arrival of the frigate which was to convey the consul to his destination. At length, learning that this vessel had been injured by a storm, he resolved to pass the winter in Spain, in hopes of finding another the following spring.

On his way to Madrid, he determined the geographical position of several important parts, and ascertained the height of the central plain of Castile. In March 1799 he was presented at the court of Aranjuez, and graciously received by the king, to whom he explained the motives which induced him to undertake a voyage to the New Continent. Being seconded in his application by the representations of an enlightened minister, Don Mariano Luis de

Urquijo, he to his great joy obtained leave to visit and explore, without impediment or restriction, all the Spanish territories in America. The impatience of the travellers to take advantage of the permission thus granted did not allow them to bestow much time upon preparations; and about the middle of May they left Madrid, crossed part of Old Castile, Leon, and Galicia, and betook themselves to Corunna, whence they were to sail for the island of Cuba.

According to the observations made by our travellers, the interior of Spain consists of an elevated table-land, formed of secondary deposites,—sandstone, gypsum, rock-salt, and Jura limestone. The climate of the Castiles is much colder than that of Toulon and Genoa, its mean temperature scarcely rising to 59° of Fahrenheit's thermometer. The central plain is surrounded by a low and narrow belt, in several parts of which the fan-palm, the date, the sugar-cane, the banana, and many plants common to Spain and the north of Africa vegetate, without suffering from the severity of the winter. In the space included between the parallels of thirty-six and forty degrees of north latitude the mean temperature ranges from 62·6° to 68·2° Fahrenheit, and by a concurrence of favourable circumstances this section has become the principal seat of industry and intellectual cultivation.

Ascending from the shores of the Mediterranean, towards the elevated plains of La Mancha and the Castiles, one imagines that he sees far inland, in the extended precipices, the ancient coast of the Peninsula; a circumstance which brings to mind the traditions of the Samothracians and certain historical testimonies, according to which the bursting of the waters through the Dardanelles, while it enlarged the basin of the Mediterranean, overwhelmed the southern part of Europe. The high central plain just described would, it may be presumed, resist the effects of the inundation until the escape of the waters by the strait formed between the Pillars of Hercules had gradually lowered the level of the Mediterranean, and thereby once more laid bare Upper Egypt on the one hand, and on the other, the fertile valleys of Tarragon, Valentia, and Murcia.

From Astorga to Corunna the mountains gradually rise, the secondary strata disappear by degrees, and the transition rocks which succeed announce the proximity of primitive formations. Large mountains of graywacke and graywacke-slate present themselves. In the vicinity of the latter town are granitic summits which extend to Cape Ortegal, and which might seem, with those of Brittany and Cornwall, to have once formed a chain of mountains that has been broken up and submersed. This rock is characterized by large and beautiful crystals of felspar, and contains tin-ore, which is worked with much labour and little profit by the Galicians.

On arriving at Corunna, they found the port blockaded by the English, for the purpose of interrupting the communication between the mother-country and the American colonies. The principal secretary of state had recommended them to Don Rafael Clavigo, recently appointed director-general of the maritime posts, who neglected nothing that could render their residence agreeable, and advised them to embark on board the corvette Pizarro bound for Havannah and Mexico. Instructions were given for the safe disposal of the instruments, and the captain was ordered to stop at Teneriffe so long as should be found necessary to enable the travellers to visit the port of Orotava and ascend the Peak.

During the few days of their detention, they occupied themselves in preparing the plants which they had collected, and in making sundry observations. Crossing to Ferrol they made some interesting experiments on the temperature of the sea and the decrease of heat in the successive strata of the water. The thermometer on the bank and near it was from 54° to 55·9°, while in deep water it stood at 59° or 59·5°, the air being 55°. The fact that the proximity of a sand-bank is indicated by a rapid descent of the temperature of the sea at its surface, is of great importance for the safety of navigators; for, although the use of the thermometer ought not to supersede that of the lead, variations of temperature indicative of danger may be perceived by it long before the vessel reaches the shoal. A heavy swell from the north-west rendered it impossible to continue their experiments. It was produced by a storm at sea, and obliged the English vessels to retire from the coast,—a circumstance which induced our travellers speedily to embark their instruments and baggage, although they were prevented from sailing by a high westerly wind that continued for several days.

CHAPTER II.

Voyage from Corunna to Teneriffe.

Departure from Corunna—Currents of the Atlantic Ocean—Marine Animals—Falling Stars—Swallows—Canary Islands—Lancerota—Fucus vitifolius—Causes of the Green Colour of Plants—La Graciosa—Stratified Basalt alternating with Marl—Hyalite—Quartz Sand—Remarks on the Distance at which Mountains are visible at Sea, and the Causes by which it is modified—Landing at Teneriffe.

THE wind having come round to the north-east, the Pizarro set sail on the afternoon of the 5th of June 1799, and after working out of the narrow passage passed the Tower of Hercules, or lighthouse of Corunna, at half-past six. Towards evening the wind increased, and the sea ran high. They directed their course to the north-west, for the purpose of avoiding the English frigates which were cruising off the coast, and about nine spied the fire of a fishing-hut at Lisarga, which was the last object they beheld in the west of Europe. As they advanced, the light mingled itself with the stars which rose on the horizon. " Our eyes," says Humboldt, " remained involuntarily fixed upon it. Such impressions do not fade from the memory of those who have undertaken long voyages at an age when the emotions of the heart are in full force. How many recollections are awakened in the imagination by a luminous point, which in the middle of a dark night, appearing at intervals above the agitated waves, marks the shore of one's native land!"

They were obliged to run under courses, and proceeded at the rate of ten knots, although the vessel was not a fast sailer. At six in the morning she rolled so much that the fore topgallant-mast was carried away. On the 7th they were in the latitude of Cape Finisterre, the group of granitic rocks on which, named the Sierra de Torinona, is visible at sea to the distance of 59 miles. On the 8th, at sunset, they discovered from the mast-head an English convoy; and to avoid them they altered their course during the night. On the 9th they began to feel the effect of the great current which flows from the Azores towards the Straits of Gibraltar and the Canaries. Its direction was at first east by south; but nearer the inlet it became due east, and its force was such as, between 37° and 30° lat., sometimes to carry the vessel, in twenty-four hours, from 21 to 30 miles eastward.

Between the tropics, especially from the coast of Senegal to the Caribbean Sea, there is a stream that always flows from east to west, and which is named the Equinoctial Current. Its mean rapidity may be estimated at ten or eleven miles in twenty-four hours. This movement of the waters, which is also observed in the Pacific Ocean, having a direction contrary to that of the earth's rotation, is supposed to be connected with the latter only in so far as it changes into trade-winds those aerial currents from the poles, which, in the lower regions of the atmosphere, carry the cold air of the high latitudes towards the equator; and it is to the general impulse which these winds give to the surface of the ocean that the phenomenon in question is to be attributed. This current carries the waters of the Atlantic towards the Mosquito and Honduras coasts, from which they move northwards, and passing into the Gulf of Mexico follow the bendings of the shore from Vera Cruz to the mouth of the Rio del Norte, and from thence to the mouths of the Mississippi and the shoals at the southern extremity of Florida. After performing this circuit, it again directs itself northward, rushing with great impetuosity through the Straits of Bahama. At the end of these narrows, in the parallel of Cape Canaveral, the flow, which rushes onward like a torrent sometimes at the rate of five miles an hour, runs to the northeast. Its velocity diminishes and its breadth enlarges as it proceeds northward. Between Cape Biscayo and the Bank of Bahama the width is only 52 miles, while in $28\frac{1}{2}°$ of lat. it is 59; and in the parallel of Charlestown, opposite Cape Henlopen, it is from 138 to 173 miles, the rapidity being from three to five miles an hour where the stream is narrow, and only one mile as it advances towards the north. To the east of Boston and in the meridian of Halifax the current is nearly 276 miles broad. Here it suddenly turns towards the east; its western margin touching the extremity of the great bank of Newfoundland. From this to the Azores it continues to flow to the E. and E.S.E., still retaining part of the impulse which it had received nearly 1150 miles distant in the Straits of Florida. In the meridian of the Isles of Corvo and Flores, the most western of the Azores, it is not less than 552 miles in breadth. From the Azores it directs itself towards the Straits of Gibraltar, the island of Madeira, and the Canary Isles. To the south of Madeira we can distinctly follow its motion to the S.E. and S.S.E. bearing on the shores of Africa, between Capes Cantin and Bojador. Cape Blanco, which, next to Cape Verd, farther to the south, is the most prominent part of that coast, seems again to influence the direction of the stream; and in this parallel it mixes with the great equinoctial current as already described.

In this manner the waters of the Atlantic, between the parallels of 11° and 43°, are carried round in a continual whirlpool, which Humboldt calculates must take two years and ten months to perform its circuit of 13,118 miles. This great current is named the Gulf-stream. Off the coast of Newfoundland a branch separates from it, and runs from S.W. to N.E. towards the coasts of Europe.

From Corunna to 36° of latitude, our travellers had scarcely seen any other animals than terns (or sea-swallows) and a few dolphins; but on the 11th June they entered a zone in which the whole sea was covered with a prodigious quantity of medusæ. The vessel was almost becalmed; but the mollusca advanced towards the south-east with a rapidity equal to four times that of the current, and continued to pass nearly three quarters of an hour, after which only a few scattered individuals were seen. Among these animals they recognised the *Medusa aurita* of Baster, the *M. pelagica* of Bosc, and a third approaching in its characters to the *M. hysocella*, which is distinguished by its yellowish-brown colour, and by having its tentacula longer than the body. Several of them were four inches in diameter, and the bright reflection from their bodies contrasted pleasantly with the azure tint of the sea.

On the morning of the 13th June, in lat. 34° 33', they observed large quantities of the *Dagysa notata*, of which several had been seen among the medusæ, and which consist of little transparent gelatinous sacs, extending to 14 lines, with a diameter of 2 or 3, and open at both ends. These cylinders are longitudinally agglutinated like the cells of a honeycomb, and form strings from six to eight inches in length. They observed, after it became dark, that none of the three species of medusa which they had collected emitted light unless they were slightly shaken. When a very irritable individual is placed on a tin plate, and the latter is struck with a piece of metal, the vibrations of the tin are sufficient to make the animal shine. Sometimes, on galvanizing medusæ, the phosphorescence appears at the moment when the chain closes, although the exciters are not in direct contact with the body of the subject. The fingers, after touching it, remain luminous for two or three minutes. Wood, on being rubbed with a medusa, becomes luminous, and, after the phosphorescence has ceased, it may be rekindled by passing the dry hand over it; but when the light is a second time extinguished it cannot be reproduced.

Between the island of Madeira and the coast of Africa, they were struck by the prodigious quantity of falling stars, which continued to increase as they advanced southward. These meteors, Humboldt remarks, are more common and more luminous in certain regions of the earth than in others. He has nowhere seen them more frequent than in the vicinity of the volcanoes of Quito, and in that part of the South Sea which washes the shores of Guatimala. According to the observations of Benzenberg and Brandes, many falling stars noticed in Europe were only 63,950 yards, or a little more than 36 miles high; and one was measured, the elevation of which did not exceed 29,843 yards, or about 17 miles. In warm climates, and especially between the tropics, they often leave behind them a train which remains luminous for twelve or fifteen seconds. At other times they seem to burst, and separate into a number of sparks. They are generally much lower than in the north of Europe. These meteors can be observed only when the sky is clear; and perhaps none has ever been seen beneath a cloud. According to the observations of M. Arago, they usually follow the same course for several hours; and in this case their direction is that of the wind.

When the voyagers were 138 miles to the east of Madeira, a common swallow (*Hirundo rustica*) perched on the topsail-yard, and was caught. What could induce a bird, asks our traveller, to fly so far at this season, and in calm weather? In the expedition of Entrecasteaux, a swallow was also seen at the distance of 207 miles off Cape Blanco; but this happened about the end of October, and M. Labillardiere imagined that it had newly arrived from Europe.

The Pizarro had been ordered to touch at Lancerota, one of the Canaries, to ascertain whether the harbour of Santa Cruz in Teneriffe was blockaded by the English; and on the 16th, in the afternoon, the seamen discovered land, which proved to be that island. As they advanced, they saw first the island of Forteventura, famous for the number of camels reared upon it, and soon after the smaller one of Lobos. Spending part of the night on deck, the naturalists viewed the volcanic summits of Lancerota illumined by the moon, and enjoyed the beautiful serenity of the atmosphere. After a time, great black clouds, rising behind the volcano, shrouded at intervals the moon and the constellation of Scorpio. They observed lights carried about on the shore, probably by fishermen, and having been employed occasionally during their passage in reading some of the old Spanish voyages, these moving fires recalled to their imagination those seen on the island of Guanahani on the memorable night of the discovery of the New World.

In passing through the archipelago of small islands, situated to the north of Lancerota, they were struck by the configuration of the coasts, which resembled the banks of the Rhine near Bonn. It is a remarkable circumstance, our author observes, that, while the forms of animals and plants exhibit the greatest diversity in different climates, the rocky masses present the same appearances in both hemispheres. In the Canary Isles, as in Auvergne, in the Mittelgebirge, in Bohemia, in Mexico, and on the banks of the Ganges, the trap formation displays a symmetrical arrangement of the mountains, exhibiting truncated cones and graduated platforms.

The whole western part of Lancerota announces the character of a country recently deranged by volcanic action, every part being black, arid, and destitute of soil. The Abbé Viera relates, that in 1730 more than half of the island changed its appearance. The great volcano ravaged the most fertile and best cultivated district, and entirely destroyed nine villages. Its eruptions were preceded by an earthquake, and violent shocks continued to be felt for several years,—a phenomenon of rare occurrence, the agitation of the ground usually ceasing after a disengagement of lava or other volcanic products. The summit of the great crater is rounded, and its absolute height does not appear to be much above 1918 feet. The island of Lancerota was formerly named Titeroigotra, and at the time of the arrival of the Spaniards its inhabitants were more civilized than the other Canarians, living in houses built of hewn stone, while the Guanches of Teneriffe resided in caves. There was then a very singular institution in the island. The women had several husbands, each of whom enjoyed the prerogative belonging to the head of a family in succession, the others remaining for the time in the capacity of common domestics.*

The occurrence, between the islands of Alegranza and Montana Clara, of a singular marine production, with light-green leaves, which was brought up by the lead from a great depth, affords our author, in his narrative, an opportunity of stating some interesting facts respecting the colouring of plants. This seaweed, growing at the bottom of the ocean at a depth of 205 feet, had its vine-shaped leaves as

* A similar practice is stated by Mr Fraser, in his " Journal of a Tour through the Himala Mountains," p. 206, to occur in several of the hill provinces of India. " It is usual all over the country for the future husband to purchase his wife from her parents; and the sum thus paid varies of course with the rank of the purchaser. The difficulty of raising this sum, and the alleged expense of maintaining women, may in part account for, if it cannot excuse, a most disgusting usage, which is universal over the country. Three or four or more brothers marry and cohabit with one woman, who is the wife of all. They are unable to raise the requisite sum individually, and thus club their shares, and buy this one common spouse."

green as those of our gramineæ. According to Bouguer's experiments, light is weakened after a passage of 192 feet, in the proportion of 1 to 1477·8. At the depth of 205, this fucus could only have had light equal to half of that supplied by a candle seen at the distance of a foot. The germs of several of the liliaceæ, the embryo of the mallows and other families, the branches of some subterranean plants, and vegetables transported into mines in which the air contains hydrogen or a great quantity of azote, become green without light. From these facts one might be induced to think that the existence of carburet of iron, which gives the green colour to the parenchmay of plants, is not dependent upon the presence of the solar rays only. Turner and many other botanists are of opinion that most of the seaweeds which we find floating on the ocean, and which in certain parts of the Atlantic present the appearance of a vast inundated meadow, grow originally at the bottom of the sea, and are torn off by the waves. If this opinion be correct, the family of marine algæ presents great difficulties to those physiologists who persist in thinking that, in all cases, the absence of light must produce blanching.

The captain, having mistaken a basaltic rock for a castle, saluted it, and sent one of the officers to inquire if the English were cruising in those parts. Our travellers took advantage of the boat to examine the land, which they had regarded as a prolongation of the coasts of Lancerota, but which turned out to be the small island of La Graciosa. "Nothing," says Humboldt, " can express the emotion a naturalist feels when for the first time he lands in a place which is not European. The attention is fixed upon so many objects, that one can hardly give an account of the impressions which he receives. At every step he imagines that he finds a new production; and, in the midst of this agitation, he often does not recognise those which are most common in our botanical gardens and museums." A fisherman, who, having been frightened by the firing, had fled from them, but whom the sailors overtook, stated that no vessels had been seen for several weeks. The rocks of this small island were of basalt and marl, destitute of trees or shrubs, in most places without a trace of soil, and but scantily crusted with lichens.

The basalts are not columnar, but arranged in strata from 10 to 16 inches thick, and incline to the north-west at an angle of 80 degrees, alternating with marl. Some of these strata are compact, and contain large crystals of foliated olivine, often porous, with oblong cavities, from two to eight lines in diameter, which are coated with calcedony, and enclose fragments of compact basalt. The marl, which alternates more than a hundred times with the trap, is of a yellowish colour, extremely friable, very tenacious internally, and often divided into regular prisms like those of basalt. It contains much lime, and effervesces strongly with muriatic acid. The travellers had not time to reach the summit of a hill, the base of which was formed of clay, with layers of basalt resting on it, precisely as in the Schneibenberger Huegel of Saxony. These rocks were covered with hyalite, of which they procured several fine specimens, leaving masses eight or ten inches square untouched.

On the shore there were two kinds of sand, the one black and basaltic, the other white and quartzy.

Exposed to the sun's rays the thermometer rose in the former to 124·2°, and in the latter to 104°; while in the shade the temperature of the air was 81·5°, being 14° higher than the sea air. The quartzy sand contains fragments of felspar. Pieces of granite have been observed at Teneriffe; and the island of Gomera, according to M. Broussonet, contains a nucleus of mica-slate. From these facts Humboldt infers that, in the Canaries, as in the Andes of Quito, in Auvergne, Greece, and most parts of the globe, the subterranean fires have made their way through primitive rocks.

Having re-embarked, they hoisted sail, and endeavoured to get out again by the strait which separates Alegranza from Montana Clara; but, the wind having fallen, the currents drove them close upon a rock marked in old charts by the name of Infierno, and in modern ones under that of Roca del Oeste,—a basaltic mass which has probably been raised by volcanic agency. Tacking during the night between Montana Clara and this islet, they were several times in great danger among shelves towards which they were drawn by the motion of the water; but the wind freshening in the morning, they succeeded in passing the channel, and sailed along the coasts of Lancerota, Lobos, and Forteventura.

The haziness of the atmosphere prevented them from seeing the Peak of Teneriffe during the whole of their passage from Lancerota; but our traveller, in his narrative, states the following interesting circumstances relative to the distance at which mountains may be seen. If the height of the Peak, he says, is 12,182 feet, as indicated by the last trigonometrical measurement of Borda, its summit ought to be visible at the distance of 148 miles, supposing the eye at the level of the ocean, and the refraction equal to 0·079 of the distance. Navigators who frequent these latitudes find that the Peaks of Teneriffe and the Azores are sometimes observed at very great distances, while at other times they cannot be seen when the interval is considerably less, although the sky is clear. Such circumstances are of importance to navigators, who, in returning to Europe, impatiently wait for a sight of these mountains to rectify their longitude. The constitution of the atmosphere has a great influence on the visibility of distant objects, the transparency of the air being much increased when a certain quantity of water is uniformly diffused through it.

It is not surprising that the Peak of Teneriffe should be less frequently visible at a great distance than the tops of the Andes, not being like them invested with perpetual snow. The Sugar-loaf which constitutes the summit of the former, no doubt reflects a great degree of light, on account of the white colour of the pumice with which it is covered; but its height does not form a twentieth part of the total elevation, and the sides of the volcano are coated with blocks of dark-coloured lava, or with luxuriant vegetation, the masses of which reflect little light, the leaves of the trees being separated by shadows of greater extent than the illuminated parts.

Hence the Peak of Teneriffe is to be referred to the class of mountains which are seen at great distances only in what Bouguer calls a negative manner, or because they intercept the light transmitted from the extreme limits of the atmosphere; and we perceive their existence only by means of the difference of intensity that subsists between the light which surrounds them, and that reflected by the particles of air placed between the object of vision and the observer. In receding from Teneriffe, the Sugarloaf is long seen in a positive manner, as it reflects a whitish light, and detaches itself clearly from the sky; but as this terminal cone is only 512 feet high, by 256 in breadth at its summit, it has been questioned whether it can be visible beyond the distance of 138 miles. If it be admitted that the mean breadth of the Sugar-loaf is $639\frac{1}{2}$ feet, it will still subtend, at the distance now named, an angle of more than three minutes, which is enough to render it visible; and were the height of the cone greatly to exceed its basis, the angle might be still less, and the mass yet make an impression on our organs; for it has been proved by micrometrical observations, that the limit of vision is one minute only when the dimensions of objects are the same in all directions.

As the visibility of an object, which detaches itself from the sky of a brown colour, depends on the quantities of light the eye meets in two lines, of which one ends at the mountain and the other is prolonged to the surface of the aerial ocean, it follows that the farther we remove from the object, the less also becomes the difference between the light of the surrounding atmosphere and that of the strata of air placed before the mountain. For this reason, when summits of low elevation begin to appear above the horizon, they are of a darker tint than those more elevated ones which we discover at very great distances. In like manner, the visibility of mountains which are only negatively perceived, does not depend solely upon the state of the low regions of the air, to which our meteorological observations are confined, but also upon its transparency and physical constitution in the most elevated parts; for the image is more distinctly detached, the more intense the aerial light which comes from the limits of the atmosphere has originally been, or the less it has lost in its passage. This in a certain degree accounts for the circumstance, that the Peak is sometimes visible and sometimes invisible to navigators who are equally distant from it, when the state of the thermometer and hygrometer is precisely the same in the lower stratum of air. It is even probable, that the chance of perceiving this volcano would not be greater, were the cone equal, as in Vesuvius, to a fourth part of the whole height. The ashes spread upon its surface do not reflect so much light as the snow with which the summits of the Andes are covered, but, on the contrary, make the mountain, when seen from a great distance, become more obscurely detached, and assume a brown tint. They contribute, as it were, to equalise the portions of aerial light, the variable difference of which renders the object more or less distinctly visible. Bare calcareous mountains, summits covered with granitic sand, and the elevated savannahs of the Andes, which are of a bright yellow colour, are more clearly seen at small distances than objects that are perceived only in a negative manner; but theory points out a limit beyond which the latter are more distinctly detached from the azure vault of the sky.

The aerial light projected on the tops of hills increases the visibility of those which are seen positively, but diminishes that of such as are detached with a brown colour. Bouguer, proceeding on theoretical data, has found that mountains which are seen negatively cannot be perceived at distances exceeding 121 miles; but experience goes against this conclusion. The Peak of Teneriffe has often been observed at the distance of 124, 131, and even 138 miles; and the summit of Mowna-Roa in the Sandwich Isles, which is probably 16,000 feet high, has been seen, at a period when it was destitute of snow, skirting the horizon from a distance of 183 miles. This is the most striking example yet known of the visibility of high land, and is the more remarkable that the object was negatively seen.

The atmosphere continuing hazy, the navigators did not discover the island of Grand Canary, notwithstanding its height, until the evening of the 18th June. On the following day they saw the point of Naga, but the Peak of Teneriffe still remained invisible. After repeatedly sounding, on account of the thickness of the mist, they anchored in the road of Santa Cruz, when at the moment they began to salute the place the fog instantaneously dispersed, and the Peak of Teyde, illuminated by the first rays of the sun, appeared in a break above the clouds. Our travellers betook themselves to the bow of the vessel to enjoy the majestic spectacle, when, at the very moment, four English ships were seen close astern. The anchor was immediately got up, and the Pizarro stood in as close as possible, to place herself under the protection of the fort.

While waiting the governor's permission to land, Humboldt employed the time in making observations for determining the longitude of the mole of

Santa Cruz and the dip of the needle. Berthoud's chronometer gave 18° 33′ 10″, the accuracy of which result, although differing from the longitude assigned by Cook and others, was afterwards confirmed by Krusenstern, who found that port 16° 12′ 45″ west of Greenwich, and consequently 18° 33′ 0″ west of Paris. The dip of the magnetic needle was 62° 24′, although it varied considerably in different places along the shore. After undergoing the fatigue of answering the numberless questions proposed by persons who visited them on board, our travellers were at length permitted to land.

CHAPTER III.

Island of Teneriffe.

Santa Cruz—Villa de la Laguna—Guanches—Present Inhabitants of Teneriffe—Climate—Scenery of the Coast—Orotava—Dragon-tree—Ascent of the Peak—Its Geological Character—Eruptions —Zones of Vegetation—Fires of St John.

SANTA CRUZ, the Anaja of the Guanches, which is a neat town with a population of 8000 persons, may be considered as a great caravansera situated on the road to America and India, and has consequently been often described. The recommendations of the court of Madrid procured for our travellers the most satisfactory reception in the Canaries. The captain-general gave permission to examine the island, and Colonel Armiaga, who commanded a regiment of infantry, extended his hospitality to them, and showed the most polite attention. In his garden they admired the banana, the papaw, and other plants cultivated in the open air, which they had before seen only in hothouses.

In the evening they made a botanical excursion towards the fort of Passo Alto, along the basaltic rocks which close the promontory of Naga, but had little success, as the drought and dust had in a manner destroyed the vegetation. The *Cacalia kleinia, Euphorbia canariensis,* and other succulent plants, which derive their nourishment more from the air than from the soil, reminded them by their aspect that the Canaries belong to Africa, and even to the most arid part of that continent.

The captain of the Pizarro having apprized them that, on account of the blockade by the English, they ought not to reckon upon a longer stay than four or five days, they hastened to set out for the port of Orotava, where they might find guides for the ascent of the Peak; and on the 20th, before sunrise, they were on the way to Villa de la Laguna, which is 2238 feet higher than the port of Santa Cruz. The road to this place is on the right of a torrent which, in the rainy season, forms beautiful falls. Near the town they met with some white camels, employed in transporting merchandise. These animals, as well as horses, were introduced into the Canary Islands in the fifteenth century by the Norman conquerors, and were unknown to the Guanches. Camels are more abundant in Lancerota and Forteventura, which are nearer the continent, than at Teneriffe, where they very seldom propagate.

The hill on which the Villa de la Laguna stands belongs to the series of basaltic mountains, which forms a girdle around the Peak, and is independent of the newer volcanic rocks. The basalt on which the travellers walked was blackish-brown, compact, and partially decomposed. They found in it hornblende, olivine, and transparent pyroxene, with lamellar fracture, of an olive-green tint, and often crystallized in six-sided prisms. The rock of Laguna is not columnar, but divided into thin beds, inclined at an angle of from 30° to 48°, and has no appearance of having been formed by a current of lava from the Peak. Some arborescent Euphorbiæ, Cacalia kleinia, and Cacti, were the only plants observed on these parched acclivities. The mules slipped at every step on the inclined surfaces of the rock, although traces of an old road were observable, which, with the numerous other indications that occur in these colonies, afford evidence of the activity displayed by the Spanish nation in the sixteenth century.

The heat of Santa Cruz, which is suffocating, is in a great measure to be attributed to the reverberation of the rocks in its vicinity; but as the travellers approached Laguna they became sensible of a very pleasant diminution of temperature. In fact, the perpetual coolness which exists here renders it a delightful residence. It is situated in a small plain, surrounded by gardens, and commanded by a hill crowned with the laurel, the myrtle, and the arbutus. The rain, in collecting, forms from time to time a kind of large pool or marsh, which has induced travellers to describe the capital of Teneriffe as situated on the margin of a lake. The town, which was deprived of its opulence in consequence of the port of Garachico having been destroyed by the lateral eruptions of the volcano, has only 9000 inhabitants, of which about 400 are monks. It is surrounded by numerous windmills for corn. Humboldt observes, that the cereal grasses were known to the original inhabitants, and that parched barley flour and goat's milk formed their principal meals. This food tends to show that they were connected with the nations of the Old Continent, perhaps even with those of the Caucasian race, and not with the inhabitants of the New World, who, previous to the arrival of the Europeans among them, had no knowledge of grain, milk, or cheese.

The Canary Islands were originally inhabited by a people famed for their tall stature, and known by the name of Guanches. They have now entirely disappeared under the oppression of a more powerful and more enlightened race, which, assuming the superiority supposed to be sanctioned by civilisation and the profession of the Christian faith, disposed of the natives in a manner little accordant with the character of a true follower of the cross. The archipelago of the Canaries was divided into small states hostile to each other; and in the fifteenth century, the Spaniards and Portuguese made voyages to these islands for slaves, as the Europeans have latterly been accustomed to do to the coast of Guinea. One Guanche then became the property of another, who sold him to the dealers; while many, rather than become slaves, killed their children and themselves. The natives had been greatly reduced in this manner, when Alonzo de Lugo completed their subjugation. The residue of that unhappy people perished by a terrible pestilence, which was supposed to have originated from the bodies left exposed by the Spaniards after the battle of Laguna. At the present day, no individual of pure blood exists in these islands, where all that remains of the aborigines are certain mummies, reduced to an extraordinary degree of desiccation, and found in the sepulchral caverns which are cut in the rock on the eastern slope of the Peak. These skeletons contain remains of aromatic plants, especially the *Chenopodium ambrosioides,* and are often decorated with

CLIMATE OF TENERIFFE. 45

small laces, to which are suspended little cakes of baked earth.

The people who succeeded the Guanches were descended from the Spaniards and Normans. The present inhabitants are described by our author as being of a moral and religious character, but of a roving and enterprising disposition, and less industrious at home than abroad. The population in 1790 was 174,000. The produce of the several islands consists chiefly of wheat, barley, maize, potatoes, wine, a great variety of fruits, sugar, and other articles of food; but the lower orders are frequently obliged to have recourse to the roots of a species of fern. The principal objects of commerce are wine, brandy, archil (a kind of lichen used as a dye), and soda.

Teneriffe has been praised for the salubrity of its climate. The ground of the Canary Islands rises gradually to a great height, and presents, on a small scale, the temperature of every zone, from the intense heat of Africa to the cold of the Alpine regions; so that a person may have the benefit of whatever climate best suits his temperament or disease. A similar variety exists as to the vegetation; and no country seemed to our travellers more fitted to dissipate melancholy, and restore peace to an agitated mind, than Teneriffe and Madeira, where the natural beauty of the situation, and the salubrity of the air, conspire to quiet the anxieties of the spirit and invigorate the body, while the feelings are not harassed by the revolting sight of slavery, which exists in almost all the European colonies.

In winter the climate of Laguna is excessively foggy, and the inhabitants often complain of cold, although snow never falls. The lowest height at which it occurs annually in Teneriffe has not been ascertained; but it has been seen in a place lying above Esperanza de la Laguna, close to the town of that name, in the gardens of which the breadfruit-tree (*Artocarpus incisa*), introduced by M. Broussonet, has been naturalized. In connexion with this subject, Humboldt remarks, that in hot countries the plants are so vigorous that they can bear a greater degree of frost than might be expected, provided it be of short duration. The banana is cultivated in Cuba, in places where the thermometer sometimes descends to very near the freezing-point; and in Spain and Italy, orange and date trees do not perish, although the cold may be two degrees below zero. Trees growing in a fertile soil are remarked by cultivators to be less delicate, and less affected by changes of temperature, than those planted in land that affords little nutriment.

From Laguna to the port of Orotava, and the western coast of Teneriffe, the route is at first over a hilly country covered by a black argillaceous soil. The subjacent rock is concealed by layers of ferruginous earth; but in some of the ravines are seen columnar basalts, with recent conglomerates, resembling volcanic tufas lying over them, which contain fragments of the former, and also, as is asserted, marine petrifactions. This delightful country, of which travellers of all nations speak with enthusiasm, is entered by the valley of Tacoronte, and presents scenes of unrivalled beauty. The seashore is ornamented with palms of the date and cocoa species. Farther up, groups of musæ and dragon-trees present themselves. The declivities are covered with vines. Orange-trees, myrtles, and cypresses, surround the chapels that have been raised on the little hills. The lands are separated by enclosures formed of the agave and cactus. Multitudes of cryptogamic plants, especially ferns, cover the walls. In winter, while the volcano is wrapped in snow, there is continued spring in this beautiful district; and in summer, towards evening, the sea-breezes diffuse a gentle coolness over it. From Tegueste and Tacoronte to the village of San Juan de la Rambla, the coast is cultivated like a garden, and might be compared to the neighbourhood of Capua or Valentia; but the western part of Teneriffe is much more beautiful, on account of the proximity of the Peak, the sight of which has a most imposing effect, and excites the imagination to penetrate into the mysterious source of volcanic action. For thousands of years no light has been observed at the summit of the mountain, and yet enormous lateral eruptions, the last of which happened in 1798, prove the activity of a fire which is far from being extinct. There is, besides, something melancholy in the sight of a crater placed in the midst of a fertile and highly-cultivated country.

Pursuing their course to the port of Orotava, the travellers passed the beautiful hamlets of Matanza and Vittoria (slaughter and victory),—names which occur together in all the Spanish colonies, and present a disagreeable contrast to the feelings of peace and quiet which these countries inspire. On their way they visited a botanic garden at Durasno, where they found M. Le Gros, the French vice-consul, who subsequently served as an excellent guide to the Peak. The idea of forming such an establishment at Teneriffe originated with the Marquis de Nava,

Dragon-tree of Orotava.

who thought that the Canary Islands afford the most suitable place for naturalizing the plants of the East and West Indies, previous to their introduction to Europe. They arrived very late at the port, and next morning commenced their journey to the Peak, accompanied by M. Le Gros, M. Lalande, secretary of the French consulate at Santa Cruz, the English gardener of Durasno, and a number of guides.

Orotava, the Taoro of the Guanches, is situated on a very steep declivity, and has a pleasant aspect when viewed from a distance, although the houses, when seen at hand, have a gloomy appearance. One of the most remarkable objects in this place is the dragon-tree in the garden of M. Franqui, of which an engraving is here presented, and which our travellers found to be about 60 feet high, with a circumference of 48 feet near the roots. The trunk divides into a great number of branches, which rise in the form of a candelabrum, and are terminated by tufts of leaves. This tree is said to have been revered by the Guanches as the ash of Ephesus was by the Greeks; and in 1402, at the time of the first expedition of Bethencour, was as large and as hollow as our travellers found it. As the species is of very slow growth, the age of this individual must be great. It is singular, that the dragon-tree should have been cultivated in these islands at so early a period, it being a native of India, and nowhere occurring on the African continent.

Leaving Orotava they passed by a narrow and stony path through a beautiful wood of chestnuts to a place covered with brambles, laurels, and arborescent heaths, where, under a solitary pine, known by the name of Pino del Dornajito, they procured a supply of water. From this place to the crater they continued to ascend without crossing a single valley, passing over several regions distinguished by their peculiar vegetation, and rested during part of the night in a very elevated position, where they suffered severely from the cold. About three in the morning they began to climb the Sugar-loaf, or small terminal cone, by the dull light of fir-torches, and examined a small subterranean glacier or cave, whence the towns below are supplied with ice throughout the summer.

In the twilight they observed a phenomenon not unusual on high mountains,—a stratum of white clouds spread out beneath, concealing the face of the ocean, and presenting the appearance of a vast plain covered with snow. Soon afterwards another very curious sight occurred, namely, the semblance of small rockets thrown into the air, and which they at first imagined to be a certain indication of some new eruption of the great volcano of Lancerota. But the illusion soon ceased, and they found that the luminous points were only the images of stars magnified and refracted by the vapours. They remained motionless at intervals, then rose perpendicularly, descended sidewise, and returned to their original position. After three hours' march over an extremely rugged tract, the travellers reached a small plain called La Rambleta, from the centre of which rises the Piton or Sugar-loaf. The slope of this cone, covered with volcanic ashes and pumice, is so steep that it would have been almost impossible to reach the summit, had they not ascended by an old current of lava, which had in some measure resisted the action of the atmosphere.

On attaining the top of this steep, they found the crater surrounded by a wall of compact lava, in which, however, there was a breach affording a passage to the bottom of the funnel or caldera, the greatest diameter of which at the mouth seemed to be 320 feet. There were no large openings in the crater; but aqueous vapours were emitted by some of the crevices, in which heat was perceptible. In fact, the volcano has not been active at the summit for thousands of years, its eruptions having been from the sides, and the depth of the crater is only about 106 feet. After examining the objects that presented themselves in this elevated spot and enjoying the vast prospect, the travellers commenced their descent, and towards evening reached the port of Orotava.

The Peak of Teneriffe forms a pyramidal mass, having a circumference at the base of more than 115,110 yards, and a height of 12,176 feet.* Two-thirds of the mass are covered with vegetation the remaining part being steril, and occupying about ten square leagues of surface. The cone is very small in proportion to the size of the mountain, it having a height of only 537 feet, or $\frac{1}{23}$ of the whole. The lower part of the island is composed of basalt and other igneous rocks of ancient formation, and is separated from the more recent lavas and the products of the present volcano by strata of tufa, puzzolana, and clay. The first that occur in ascending the Peak are of a black colour, altered by decomposition, and sometimes porous. Their basis is wacke, and has usually an irregular, but sometimes a conchoidal fracture. They are divided into very thin layers, and contain olivine, magnetic iron, and augite. On the first elevated plain, that of Retama, the basaltic deposites disappear beneath heaps of ashes and pumice. Beyond this are lavas, with a basis of pitch-stone and obsidian, of a blackish-brown or deep olive-green colour, and containing crystals of felspar, which are

---

* Various measurements have been made of the height of the Peak of Teneriffe; but Humboldt, after enumerating fourteen, states that the following alone can be considered as deserving of confidence:

>   Borda's, by trigonometry,..........1905 toises.
>   Borda's, by the barometer,........1976
>   Lamanou's, by the same,..........1902
>   Cordier's, by the same,............1920

The average of these four observations makes the height 1926 toises; but if the barometric measurement of Borda be rejected, as liable to objections particularly stated by our author, the mean of the remaining measurement is 1909 toises, or 12,208 English feet. It is seen above, that the height adopted by Humboldt is 1904 toises, or 12,176 English feet.

seldom vitreous. In the middle of the Malpays or second platform are found, amongst the glassy kinds, blocks of greenish-gray clinkstone or porphyry-slate. Obsidian of several varieties is exceedingly abundant on the Peak, as well as pumice, the latter being generally of a white colour; and the crater contains an enormous quantity of sulphur.

The oldest written testimony, in regard to the activity of the volcano, dates at the beginning of the sixteenth century, and is contained in the narrative of Aloysio Cadamusto, who landed in the Canaries in 1505. In 1558, 1646, and 1677, eruptions took place in the isle of Palma; and on the 31st December 1704, the Peak of Teneriffe exhibited a lateral burst, preceded by tremendous earthquakes. On the 5th of January 1705, another opening occurred, the lavas produced by which filled the whole valley of Fasnia. This aperture closed on the 13th of January; but on the 2d of February, a third formed in the Cannada de Arafo, the stream from which divided into three currents. On the 5th May 1706, another eruption supervened, which destroyed the populous and opulent city of Garachico. In 1730, on the 1st September, the island of Lancerota was violently convulsed; and on the 9th June 1798, the Peak emitted a great quantity of matter, which continued to run three months and six days.

The island of Teneriffe presents five zones of vegetation, arranged in stages one above another, and occupying a perpendicular height of 3730 yards.

1. The *Region of Vines* extends from the shores to an elevation varying from 430 to 640 yards, and is the only part carefully cultivated. It exhibits various species of arborescent Euphorbiæ, Mesembryan-

thema, the Cacalia Kleinia, the Dracœna, and other plants, whose naked and tortuous trunks, succulent leaves, and bluish-green tints, constitute features distinctive of the vegetation of Africa. In this zone are raised the date-tree, the plantain, the sugar-cane, the Indian-fig, the arum colocasia, the olive, the fruit-trees of Europe, the vine, and wheat.

2. The *Region of Laurels* is that which forms the woody part of Teneriffe, where the surface of the ground is always verdant, being plentifully watered by springs. Four kinds of laurel, an oak, a wild olive, two species of iron-tree, the arbutus callicarpa, and other evergreens adorn this zone. The trunks are covered by the ivy of the Canaries and various twining shrubs, and the woods are filled with numerous species of fern. The hypericum, and other showy plants, enrich with their beautiful flowers the verdant carpet of moss and grass.

3. The *Region of Pines*, which commences at the height of 1920 yards, and has a breadth of 850, is characterized by a vast forest of trees, resembling the Scotch fir, intermixed with juniper.

4. The fourth zone is remarkable chiefly for the profusion of *retama*, a species of broom, which forms oases in the midst of a wide sea of ashes. It grows to the height of nine or ten feet, is ornamented with fragrant flowers, and furnishes food to the goats, which have run wild on the Peak from time immemorial.

5. The fifth zone is the *Region of the Grasses*, in which some species of these supply a scanty covering to the heaps of pumice, obsidian, and lava. A few cryptogamic plants are observed higher; but the summit is entirely destitute of vegetation.

Thus the whole island may be considered as a forest of laurels, arbutuses, and pines, of which the external margin only has been in some measure cleared, while the central part consists of a rocky and steril soil, unfit even for pasturage.

The following day was passed by our travellers in visiting the neighbourhood of Orotava, and enjoying an agreeable company at Mr Cologan's. On the eve of St John, they were present at a pastoral fête in the garden of Mr Little, who had reduced to cultivation a hill covered with volcanic substances, from which there is a magnificent view of the Peak, the villages along the coast, and the isle of Palma. Early in the evening, the volcano suddenly exhibited a most extraordinary spectacle, the shepherds having, in conformity to ancient custom, lighted the fires of St John; the scattered masses of which, with the columns of smoke driven by the wind, formed a fine contrast to the deep verdure of the woods that covered the sides of the mountain, while the silence of nature was broken at intervals by the shouts of joy which came from afar.

CHAPTER IV.

Passage from Teneriffe to Cumana.

Departure from Santa Cruz—Floating Seaweeds—Flying-fish—Stars—Malignant Fever—Island of Tobago—Death of a Passenger—Island of Coche—Port of Cumana—Observations made during the Voyage; Temperature of the Air; Temperature of the Sea; Hygrometrical State of the Air; Colour of the Sky and Ocean.

HAVING sailed from Santa Cruz on the evening of the 25th of June, with a strong wind from the north-east, our travellers soon lost sight of the Canary Islands, the mountains of which were covered with reddish vapour, the Peak alone appearing at intervals in the breaks. The passage from Teneriffe to Cumana was performed in twenty days, the distance being 3106 miles.

The wind gradually subsided as they retired from the African coast. Short calms of several hours occasionally took place, which were regularly interrupted by slight squalls, accompanied by masses of dark clouds, emitting a few large drops of rain, but without thunder. To the north of the Cape Verd Islands they met with large patches of floating seaweed (*Fucus natans*), which grows on submarine rocks, from the equator to forty degrees of latitude on either side. These scattered plants, however, must not be confounded with the vast beds, said by Columbus to resemble extensive meadows, and which inspired with terror the crew of the Santa Maria. From a comparison of numerous journals it appears that there are two such fields of seaweed in the Atlantic. The largest occurs a little to the west of the meridian of Fayal, one of the Azores, between 25° and 36° of latitude. The temperature of the ocean there is between 60·8° and 68°; and the north-west winds, which blow sometimes with impetuosity, drive floating islands of those weeds into low latitudes, as far as the parallels of 24° and even 20°. Vessels returning to Europe from Monte Video, or the Cape of Good Hope, pass through this marine meadow, which the Spanish pilots consider as lying half-way between the West Indies and the Canaries. The other section is not so well known, and occupies a smaller space between lat. 22° and 26° of N., two hundred and seventy-six miles eastward of the Bahama Islands.

Although a species of seaweed, the *Laminaria pyrifera* of Lamouroux, has been observed with stems 850 feet in length, and although the growth of these plants is exceedingly rapid, it is yet certain that in those seas the fuci are not fixed to the bottom, but float in detached parcels at the surface. In this state vegetation, it is obvious, cannot continue longer than in the branch of a tree separated from the trunk; and it may therefore be supposed, that floating masses of these weeds occurring for ages in the same position owe their origin to submarine rocks, which continually supply what has been carried off by the equinoctial currents. But the causes by which these plants are detached are not yet sufficiently known, although the author just named has shown that fuci in general separate with great facility after the period of fructification.

Beyond 22° of latitude they found the surface of the sea covered with flying-fish (*Exocetus volitans*), which sprung into the air to a height of twelve, fifteen, and even eighteen feet, and sometimes fell on the deck. The great size of the swimming-bladder in these animals, being two-thirds the length of their body, as well as that of the pectoral fins, enable them to traverse in the air a space of twenty-four feet horizontal distance before falling again into the water. They are incessantly pursued by dolphins while under the surface, and when flying are attacked by frigate-birds and other predatory species. Yet it does not seem that they leap into the atmosphere merely to avoid their enemies; for, like swallows, they move by thousands in a right line, and always in a direction opposite to that of the waves. The air contained in the swimming-bladder had been supposed to be pure oxygen; but Humboldt found it to consist of ninety-four parts of azote, four of oxygen, and two of carbonic acid.

On the 1st July they met with the wreck of a vessel, and on the 3d and 4th crossed that part of the ocean where the charts indicate the bank of the Maal-Stroom, which, however, is of very doubtful existence. As they approached this imaginary whirlpool they observed no other motion in the waters than that produced by a current bearing to the north-west.

From the time when they entered the torrid zone (the 27th June) they never ceased to admire the nocturnal beauty of the southern sky, which gradually disclosed new constellations to their view. "One experiences an indescribable sensation," says Humboldt, "when, as he approaches the equator, and especially in passing from the one hemisphere to the other, he sees the stars with which he has been familiar from infancy gradually approach the horizon and finally disappear. Nothing impresses more vividly on the mind of the traveller the vast distance to which he has been removed from his native country than the sight of a new firmament. The grouping of the larger stars, the scattered nebulæ rivalling in lustre the milky-way, and spaces remarkable for their extreme darkness, give the southern heavens a peculiar aspect. The sight even strikes the imagination of those who, although ignorant of astronomy, find pleasure in contemplating the celestial vault, as one admires a fine landscape or a majestic site. Without being a botanist, the traveller knows the torrid zone by the mere sight of its vegetation; and without the possession of astronomical knowledge perceives that he is not in Europe, when he sees rising in the horizon the great constellation of the Ship, or the phosphorescent clouds of Magellan. In the equinoctial regions, the earth, the sky, and all their garniture, assume an exotic character."

The intertropical seas being usually smooth, and the vessel being impelled by the gentle breezes of the trade-wind, the passage from the Cape Verd Islands to Cumana was as pleasant as could be desired; but as they approached the West Indies a malignant fever disclosed itself on board. The ship was very much encumbered between decks, and from the time they passed the tropic the thermometer stood from 93° to 96·8°. Two sailors, several passengers, two negroes from the coast of Guinea, and a mulatto child, were attacked. An ignorant

Galician surgeon ordered bleedings, to obviate the "heat and corruption of the blood;" but little exertion had been made in attempting to diminish the danger of infection, and there was not an ounce of bark on board. A sailor, who had been on the point of expiring, recovered his health in a singular manner. His hammock having been so hung that the sacrament could not be administered to him, he was removed to an airy place near the hatchway, and left there, his death being expected every moment. The transition from a hot and stagnant to a fresher and purer atmosphere gradually restored him, and his recovery furnished the doctor with an additional proof of the necessity of bleeding and evacuation,—a treatment of which the fatal effects soon became perceptible.

On the 13th, early in the morning, very high land was seen. The wind blew hard, the sea was rough, large drops of rain fell at intervals, and there was every appearance of stormy weather. Considerable doubt existed as to the latitude and longitude, which was, however, removed by observations made by our travellers, and the appearance of the island of Tobago. This little island is a heap of rocks, the dazzling whiteness of which forms an agreeable contrast with the verdure of the scattered tufts of trees upon it. The mountains are crowned with very tall opuntiæ, which alone are enough to apprize the navigator that he has arrived on an American coast.

After doubling the north cape of Tobago and the point of St Giles, they discovered from the mast-head what they regarded as a hostile squadron; which, however, turned out to be only a group of rocks. Crossing the shoal which joins the former island to

Grenada, they found that, although the colour of the sea was not visibly changed, the thermometer indicated a temperature several degrees lower than that of the neighbouring parts. The wind diminished after sunset, and the clouds dispersed as the moon reached the zenith. Numerous falling-stars were seen on this and the following nights.

On the 14th, at sunrise, they were in sight of the Bocca del Drago, and distinguished the island of Chacachacarreo. When 17 miles distant from the coast, they experienced, near Punta de la Baca, the effect of a current which drew the ship southward. Heaving the lead, they found from 230 to 275 feet, with a bottom of very fine green clay,—a depth much less than, according to Dampier's rule, might have been expected in the vicinity of a shore formed of very elevated and perpendicular mountains.

The disease which had broken out on board the Pizarro made rapid progress from the time they approached the coast. The thermometer kept steady at night between 71·6° and 73·4°, and during the day rose to between 75·2° and 80·6°. The determination to the head, the extreme dryness of the skin, the prostration of strength, and all the other symptoms became more alarming; but it was hoped that the sick would recover as soon they were landed on the island of St Margaret or at the port of Cumana, both celebrated for their great salubrity. This hope, however, was not entirely realized, for one of the passengers fell a victim to the distemper. He was an Asturian, nineteen years of age, the only son of a poor widow. Various circumstances combined to render the death of this young man affecting. He was of an exceedingly gentle disposition, bore the marks of great sensibility, and had left his native land against his inclination, with the view of earning an independence and assisting his reluctant mother, under the protection of a rich relation, who resided in the island of Cuba. From the commencement of his illness he had fallen into a lethargic state, interrupted by accessions of delirium, and on the third day expired. Another Asturian, who was still younger, did not leave the bed of his dying friend for a moment, and yet escaped the disease. He had intended to accompany his countryman to Cuba, to be introduced by him to the house of his relative, on whom all their hopes rested ; and it was distressing to see his deep sorrow, and to hear him curse the fatal counsels which had thrown him into a foreign climate, where he found himself alone and destitute.

"We were assembled on the deck," says our eloquent author, "absorbed in melancholy reflections. It was no longer doubtful that the fever which prevailed on board had of late assumed a fatal character. Our eyes were fixed on a mountainous and desert coast, on which the moon shone at intervals through the clouds. The sea, gently agitated, glowed with a feeble phosphoric light. No sound came on the ear save the monotonous cry of some large seabirds that seemed to be seeking the shore. A deep calm reigned in these solitary places ; but this calm of external nature accorded ill with the painful feelings which agitated us. About eight the death-bell was slowly tolled. At this doleful signal the sailors ceased from their work, and threw themselves on their knees to offer up a short prayer ; an affecting ceremony, which, while it recalls the times when the primitive Christians considered themselves as members of the same family, seems to unite men by the feeling of a common evil. In the course of the night the body of the Asturian was brought upon deck, and the priest prevailed upon them not to throw it into the sea until after sunrise, in order that he might render to it the last rites, in conformity to the practice of the Romish church. There was not an individual on board who did not feel for the fate of this young man, whom we had seen a few days before full of cheerfulness and health."

The passengers who had not been affected by the disease resolved to leave the ship at the first place where she should touch, and there wait the arrival of another packet to convey them to Cuba and Mexico. Our travellers also thought it prudent to land at Cumana, more especially as they wished not to visit New Spain until they had remained for some time on the coasts of Venezuela and Paria, and examined the beautiful plants of which Bosc and Bredemeyer collected specimens on their voyage to Terra Firma, and which Humboldt had seen in the gardens of Schönbrunn and Vienna. This resolution had a happy influence upon the direction of their journey, as will subsequently be seen, and perhaps was the occasion of securing for them the health which they enjoyed during a long residence in the equinoctial regions. They were by this means fortunate enough to pass the time when a European recently landed runs the greatest danger of being affected by the yellow fever, in the hot but very dry climate of Cumana, a city celebrated for its salubrity.

As the coast of Paria stretches to the west, in the form of perpendicular cliffs of no great height, they were long without perceiving the bold shores of the island of St Margaret, where they intended to stop for the purpose of obtaining information respecting the English cruisers. Toward eleven in the morning of the 15th, they observed a very low islet covered with sand, and destitute of any trace of culture or habitation. Cactuses rose here and there from a scanty soil which seemed to have an undulating motion, in consequence of the extraordinary refraction the solar rays undergo in passing through the stratum of air in contact with a strongly-heated surface. The deserts and sandy shores of all countries present this appearance. The aspect of this place not corresponding with the ideas which they had formed of the island of Margaretta, and the greatest perplexity existing as to their position and course, they cast anchor in shallow water, and were visited by some Guayquerias in two canoes, constructed each of the single trunk of a tree. These Indians, who were of a coppery colour and very tall, informed them that they had kept too far south, that the low islet near which they were at anchor was the island of Coche, and that Spanish vessels coming from Europe usually passed to the northward of it. The master of one of the canoes offered to remain on board as coasting pilot, and towards evening the captain set sail.

On the 16th they beheld a verdant coast of picturesque appearance; the mountains of New Andalusia bounded the southern horizon, and the city of Cumana and its castle appeared among groups of trees. They anchored in the port about nine in the morning, when the sick crawled on deck to enjoy the sight. The river was bordered with cocoa-trees more than sixty feet high,—the plain was covered with tufts of cassias, capers, and arborescent mimosas, while the pinnated leaves of the palms were conspicuous on the azure of a sky unsullied by the least trace of vapour. A dazzling light was spread along the white hills clothed with cylindrical cactuses, and over the smooth sea, the shores of which were peopled by pelicans, egrets, and flamingoes. Every thing announced the magnificence of nature in the equinoctial regions.

Before accompanying our learned friends to the city of Cumana, we may here take a glance of the physical observations made by them during the voyage, and which refer to the temperature of the air and sea, and other subjects of general interest.

Temperature of the Air.—In the basin of the northern Atlantic Ocean, between the coasts of Europe, Africa, and America, the temperature of the atmosphere exhibits a very slow increase. From Corunna to the Canary Islands, the thermometer, observed at noon and in the shade, gradually rose from 50° to 64°, and from Teneriffe to Cumana from 64° to 77°. The maximum of heat observed during the voyage did not exceed 79·9°.

The extreme slowness with which the temperature increases during a voyage from Spain to South America is highly favourable to the health of Europeans, as it gradually prepares them for the intense heat which they have to experience. It is in a great measure attributable to the evaporation of the water, augmented by the motion of the air and waves, together with the property possessed by transparent liquids of absorbing very little light at their surface. On comparing the numerous observations made by navigators, we are surprised to see that in the torrid zone, in either hemisphere, they have not found the thermometer to rise in the open sea above 93°; while in corresponding latitudes on the continents of Asia and Africa, it attains a much greater elevation. The difference between the temperature of the day and night is also less than on land.

Temperature of the Sea.—From Corunna to the mouth of the Tagus, the temperature of the sea varied little (between 59° and 60·8°) ; but from lat. 39° to 10° N., the increase was rapid and generally uniform (from 59° to 78·4°), although inequalities occurred, probably caused by currents. It is very remarkable that there is a great uniformity in the maximum of heat every where in the equinoctial waters. This maximum, which varies from 82° to 84·2°, proves that the ocean is in general warmer than the atmosphere in direct contact with it, and of which the mean temperature near the equator is from 78·8° to 80·6°.

Hygrometrical State of the Air.—During the whole of the voyage, the apparent humidity of the atmosphere indicated by the hygrometer underwent a sensible increase. In July, in lat. 13° and 14° N., Saussure's hygrometer marked at sea from 88° to 92°, in perfectly clear weather, the thermometer being at 75·2°. On the banks of the Lake of Geneva the mean humidity of the same month is only 80°, the average heat being 66·2°. On reducing these observations to a uniform temperature, we find that the real humidity in the equinoctial basin of the Atlantic Ocean is to that of the summer months at Geneva as 12 to 7. This astonishing degree of moisture in the air accounts to a great extent for the vigorous vegetation which presents itself on the coasts of South America, where so little rain falls throughout the year.

Intensity of the Colour of the Sky and Ocean.—
From the coasts of Spain and Africa to those of
South America, the azure colour of the sky increased from 13° to 23° of Saussure's cyanometer.
From the 8th to the 12th of July, in lat. $12\frac{1}{2}°$
and 14° N., the sky, although free of vapour, was of
an extraordinary paleness, the instrument indicating only 16° or 17°, although on the preceding days
it had been at 22°. The tint of the sky is generally deeper in the torrid zone than in high latitudes,
and in the same parallel it is fainter at sea than on
land. The latter circumstance may be attributed
to the quantity of aqueous vapour which is continually rising towards the higher regions of the
air from the surface of the sea. From the zenith to
the horizon, there is in all latitudes a diminution of
intensity, which follows nearly an arithmetical pro-
gression, and depends upon the moisture suspended
in the atmosphere. If the cyanometer indicate this
accumulation of vapour in the more elevated portion of the air, the seaman possesses a simpler method of judging of the state of its lower regions, by
observing the colour and figure of the solar disk at
its rising and setting. In the torrid zone, where
meteorological phenomena follow each other with
great regularity, the prognostics are more to be de-
pended upon than in northern regions. Great paleness of the setting sun, and an extraordinary dis-
figuration of its disk, almost certainly presage a
storm; and yet one can hardly conceive how the
condition of the lower strata of the air, which is
announced in this manner, can be so intimately
connected with those atmospherical changes that
take place within the space of a few hours.

Mariners are accustomed to observe the appearances of the sky more carefully than landsmen, and among ᴅthe numerous meteorological rules which pilots transmit to each other, several evince great sagacity. Prognostics are also in general less uncertain on the ocean, and especially in the equinoctial parts of it, than on land, where the inequalities of the ground interrupt the regularity of their manifestation.

Humboldt also applied the cyanometer to measure the colour of the sea. In fine calm weather, the tint was found to be equal to 33°, 38°, sometimes even 44° of the instrument, although the sky was very pale, and scarcely attained 14° or 15°. When, instead of directing the apparatus to a great extent of open sea, the observer fixes his eyes on a small part of its surface viewed through a narrow aperture, the water appears of a rich ultramarine colour. Towards evening again, when the edge of the waves, as the sun shines upon them, is of an emerald-green, the surface of the shaded side reflects a purple hue. Nothing is more striking than the rapid changes which the colour of the sea undergoes under a clear sky, in the midst of the ocean and in deep water, when it may be seen passing from indigo-blue to the deepest green, and from this to slate-gray. The blue is almost independent of the reflection of the atmosphere. The intertropical seas are in general of a deeper and purer tint than in high latitudes, and the ocean often remains blue, when, in fine weather, more than four-fifths of the sky are covered with light and scattered clouds of a white colour.

CHAPTER V.

Cumana.

Landing at Cumana—Introduction to the Governor—State of the Sick—Description of the Country and City of Cumana—Mode of Bathing in the Manzanares—Port of Cumana—Earthquakes; Their Periodicity; Connexion with the State of the Atmosphere; Gaseous Emanations; Subterranean Noises; Propagation of Shocks; Connexion between those of Cumana and the West Indies; and General Phenomena.

THE city of Cumana, the capital of New Andalusia, is a mile distant from the landing-place, and in proceeding towards it our travellers crossed a large sandy plain, which separates the suburb inhabited by the Guayqueria Indians from the seashore. The excessive heat of the atmosphere was increased by the reflection of the sun's rays from a naked soil, the thermometer immersed in which rose to 99·9°. In the little pools of salt water it remained at 86·9°, while the surface of the sea in the port generally ranges from 77·4° to 79·3°. The first plant gathered by them was the *Avicennia tomentosa,* which is remarkable for occurring also on the Malabar coast, and belongs to the small number that live' in society, like the heaths of Europe, and are seen in the torrid zone only on the shores of the ocean and the elevated platforms of the Andes.

Crossing the Indian suburb, the streets of which were 'very neat, they were conducted by the captain of the Pizarro to the governor of the province, Don Vicente Emparan, who received them with frankness; expressed his satisfaction at the resolution which they had taken of remaining for some time in New Andalusia; showed them cottons dyed with native plants and furniture made of indigenous wood; and surprised them with questions indicative of scientific attainments. On disembarking their instruments, they had the pleasure of finding that none of them had been damaged. They hired a spacious house in a situation favourable for astronomical observations, in which they enjoyed an agreeable coolness when the breeze arose, the windows being without glass, or even the paper panes which are often substituted for it at Cumana.

The passengers all left the vessel. Those who had been attacked by the fever recovered so very slowly, that some were seen a month after, who, notwithstanding the care bestowed upon them by their countrymen, were still in a state of extreme debility. The hospitality of the inhabitants of the Spanish colonies is such that the poorest stranger is sure of receiving the kindest treatment. Among the sick landed here was a negro, who soon fell into a state of insanity and died; which fact our author mentions, as a proof that persons born in the torrid zone are liable to suffer from the heat of the tropics after having resided in temperate climates. This individual, who was a robust young man, was a native of Guinea, but had lived for some years on the elevated plain of Castile.

The soil around Cumana is composed of gypsum and calcareous breccia, and is supposed at a remote period to have been covered by the sea. The neighbourhood of the city is remarkable for the woods of cactus which are spread over the arid lands. Some of these plants were thirty or forty feet high, covered with lichens, and divided into branches in the form of a candelabrum. When the large species grow in groups they form a thicket which, while it is almost impenetrable, is extremely dangerous on account of the poisonous serpents that frequent it.

The fortress of St Antonio, which is built on a calcareous hill, commands the town and forms a picturesque object to vessels entering the port. On the south-western slope of the same rock are the ruins of the castle of St Mary, from the site of which there is a fine view of the Gulf, together with the island of Margaretta and the small isles of Caraccas, Picuita, and Boracha, which present the most singular appearances from the effect of mirage.

The city of Cumana, properly speaking, occupies the ground that lies between the castle of St Antonio and the small rivers Manzanares and Santa Catalina. It has no remarkable buildings, on account of the violent earthquakes to which it is subject. The suburbs are almost as populous as the town itself, and are three in number: namely, Serritos, St Francis, and that of the Guayquerias. The latter is inhabited by a tribe of civilized Indians, who, for upwards of a century, have adopted the Castilian language. The whole population in 1802 was about eighteen or nineteen thousand.

The plains which surround the city have a parched and dusty aspect. The hill on which the fort of St Antonio stands is also bare, and composed of calcareous breccia, containing marine shells. Southward, in the distance, is a vast curtain of inaccessible mountains, also of limestone. These ridges are covered by majestic forests, extending along the sloping ground at their base to an open plain in the neighbourhood of Cumana, through which the river Manzanares winds its way to the sea, fringed with mimosas, erythrinas, ceibas, and other trees of gigantic growth.

This river, the temperature of which in the season of the floods descends as low as 71·6°, when that of the air is as high as 91°, is an inestimable benefit to the inhabitants; all of whom, even the women of the most opulent families, learn to swim. The mode of bathing is various. Our travellers frequented every evening a very respectable society in the suburb of the Guayquerias. In the beautiful moonlight chairs were placed in the water, on which were seated the ladies and gentlemen, lightly clothed. The family and the strangers passed several hours in the river, smoking cigars and chatting on the usual subjects of conversation, such as the extreme drought, the abundance of rain in the neighbouring districts, and the female luxury which prevails in Caraccas and Havannah. The company were not disturbed by the *bavas*, or small crocodiles, which are only three or four feet long, and are now extremely rare. Humboldt and his companions did not meet with any of them in the Manzanares; but they saw plenty of dolphins, which sometimes ascended the river at night, and frightened the bathers by spouting water from their nostrils.

The port of Cumana is capable of receiving all the navies of Europe; and the whole of the Gulf of Cariaco, which is forty-two miles long and from seven to nine miles broad, affords excellent anchorage. The hurricanes of the West Indies are never experienced on these coasts, where the sea is constantly smooth, or only slightly agitated by an easterly wind. The sky is often bright along the shores, while stormy clouds are seen to gather among the mountains. Thus, as at the foot of the Andes, on the western side of the continent, the extremes of clear weather and fogs, of drought and heavy rain, of absolute nakedness and perpetual verdure, present themselves on the coasts of New Andalusia.

The same analogy exists as to earthquakes, which are frequent and violent at Cumana. It is a generally-received opinion that the Gulf of Cariaco owed its existence to a rent of the continent, the remembrance of which was fresh in the minds of the natives at the time of Columbus' third voyage. In 1530, the coasts of Paria and Cumana were agitated by shocks; and towards the end of the sixteenth century, earthquakes and inundations very often occurred. On the 21st October 1766, the city of Cumana was entirely destroyed in the space of a few minutes. The earth opened in several parts of the province, and emitted sulphureous waters. During the years 1766 and 1767, the inhabitants encamped in the streets, and they did not begin to rebuild their houses until the earthquakes took place only once in four weeks. These commotions had been preceded by a drought of fifteen months, and were accompanied and followed by torrents of rain which swelled the rivers.

On the 14th December 1797, more than four-fifths of the city were again entirely destroyed. Previous to this, the shocks had been horizontal oscillations; but the shaking now felt was that of an elevation of the ground, and was attended by a subterraneous noise, like the explosion of a mine at a great depth. The most violent concussion, however, was preceded by a slight undulating motion, so that the inhabitants had time to escape into the streets; and only a few perished, who had betaken themselves for safety to the churches. Half an hour before the catastrophe, a strong smell of sulphur was experienced near the hill of the convent of St Francis; and on the same spot an internal noise, which seemed to pass from S.E. to N.W., was heard loudest. Flames appeared on the banks of the Manzanares and in the Gulf of Cariaco. In describing this frightful convulsion of nature, our author enters upon general views respecting earthquakes, of which a very brief account may be here given.

The great earthquakes which interrupt the long series of small shocks, do not appear to have any stated times at Cumana, as they have occurred at intervals of eighty, of a hundred, and sometimes even of less than thirty years; whereas, on the coasts of Peru,—at Lima, for example,—there is, without doubt, a certain degree of regularity in the periodical devastations thereby occasioned.

It has long been believed at Cumana, Acapulco, and Lima, that there exists a perceptible relation between earthquakes and the state of the atmosphere which precedes these phenomena. On the coasts of New Andalusia the people become uneasy when, in excessively hot weather and after long drought, the breeze suddenly ceases, and the sky, clear at the zenith, presents the appearance of a reddish vapour near the horizon. But these prognostics are very uncertain, and the dreaded evil has arrived in all kinds of weather.

Under the tropics the regularity of the horary variations of the barometer is not disturbed on the days when violent shocks occur. In like manner, in the temperate zone the aurora borealis does not always modify the variations of the needle, or the intensity of the magnetic forces.

When the earth is opened and agitated, gaseous emanations occasionally escape in places considerably remote from unextinguished volcanoes. At Cumana, flames and sulphureous vapours spring from the arid soil, while in other parts of the same province it throws out water and petroleum. At Riobamba, a muddy inflammable mass called *moya* issues from crevices which close again, and forms elevated heaps. Flames and smoke were also seen to proceed from the rocks of Alvidras, near Lisbon, during the earthquake of 1755, by which that city was ravaged. But in the greater number of earthquakes it is probable that no elastic fluids escape from the ground, and when gases are evolved, they more frequently accompany or follow than precede the shocks.

The subterranean noise which so frequently attends earthquakes, is generally not proportionate to the strength of the shocks. At Cumana it always precedes them, while at Quito, and for some time past at Caraccas and in the West India Islands, a noise like the discharge of a battery was heard long after the agitation had ceased. The rolling of thunder in the bowels of the earth, which continues for months, without being accompanied by the least shaking, is a very remarkable phenomenon.

In all countries subject to earthquakes the point at which the effects are greatest is considered as the source or focus of the shocks. We forget that the rapidity with which the undulations are propagated to great distances, even across the basin of the ocean, proves the centre of action to be very remote from the earth's surface. Hence it is clear that earthquakes are not restricted to certain species of rocks, as some naturalists assert, but pervade all; although sometimes, in the same rock, the upper strata seem to form an insuperable obstacle to the propagation of the motion. It is curious also, that in a district of small extent, certain formations interrupt the shocks. Thus, at Cumana, before the catastrophe of 1797, the earthquakes were felt only along the southern or calcareous coast of the Gulf of Cariaco, as far as the town of that name, while in the peninsula of Araya, and at the village of Maniquarez, the ground was not agitated. At present, however, the peninsula is as liable to earthquakes as the district around Cumana.

In New Andalusia, as in Chili and Peru, the shocks follow the line of the shore, and extend but little into the interior,—a circumstance which indicates an intimate connexion between the causes that produce earthquakes and volcanic eruptions. If the land along the coasts is most agitated because it is generally lowest, why should not the shocks be equally strong in the savannahs, which are only a few yards above the level of the sea?

The earthquakes of Cumana are connected with those of the West Indies, and are even suspected to have some relation to the volcanic phenomena of the Andes. On the 4th November 1797, the province of Quito underwent so violent a commotion that 40,000 persons were destroyed; and at the same period shocks were experienced in the Eastern Antilles, followed by an eruption of the volcano of Guadaloupe, in the end of September 1798. On the 14th December the great concussion took place at Cumana.

It has long been remarked that earthquakes extend their effects to much greater distances than volcanoes; and it is probable, as has just been mentioned, that the causes which produce the former have an intimate connexion with the latter. When seated within the verge of a burning crater, one feels the motion of the ground several seconds before each partial eruption. The phenomena of earthquakes seem strongly to indicate the action of elastic fluids endeavouring to force their way into the atmosphere. On the shores of the South Sea the concussion is almost instantaneously communicated from Chili to the Gulf of Guayaquil, over a space of 2070 miles. The shocks also appear to be so much the stronger the more distant the country is from active volcanoes; and a province is more agitated, the smaller the number of funnels by which the subterranean cavities communicate with the open air.

CHAPTER VI.

Residence at Cumana.

Lunar Halo—African Slaves—Excursion to the Peninsula of Araya —Geological Constitution of the Country—Salt-works of Araya —Indians and Mulattoes—Pearl-fishery—Maniquarez—Mexican Deer—Spring of Naphtha.

THE occupations of our travellers were much disturbed during the first weeks of their abode at Cumana by the intrusion of persons desirous of examining their astronomical and other instruments. They however determined the latitude of the great square to be 10° 27′ 52″, and its longitude 66° 30′ 2″.

On the 17th of August, a halo of the moon attracted the attention of the inhabitants, who viewed it as the presage of a violent earthquake. Coloured circles of this kind, Humboldt remarks, are much rarer in the northern than in the southern countries of Europe. They are seen more especially when the sky is clear and the weather settled. In the torrid zone they appear almost every night, and often in the space of a few minutes disappear several times. Between the latitude of 15° N. and the equator he has seen small haloes around the planet Venus, but never observed any in connexion with the fixed stars. While the halo was seen at Cumana, the hygrometer indicated great humidity, although the atmosphere was perfectly transparent. It consisted of two circles; a larger, of a whitish colour, and 44° in diameter, and a smaller, displaying all the tints of the rainbow, and 1° 43' in diameter. The intermediate space was of the deepest azure.

Part of the great square is surrounded with arcades, over which is a long wooden gallery, where slaves imported from the coast of Africa are sold. These were young men from fifteen to twenty years of age. Every morning cocoa-nut oil was given them, with which they rubbed their skin, to render it glossy. The persons who came to purchase them examined their teeth, as we do those of horses, to judge of their age and health. Yet the Spanish laws, according to our author, have never favoured the trade in African slaves, the number of whom in 1800 did not exceed 6000 in the two provinces of Cumana and Barcelona, while the whole population was estimated at 110,000.

The first excursion which our travellers made was to the peninsula of Araya. They embarked on the Manzanares, near the Indian suburb, about two in the morning of the 19th August. The night was delightfully cool. Swarms of shining insects (*Elater noctilucus*) sparkled in the air along the banks of the river. As the boat descended the stream they observed a company of negroes dancing to the music of the guitar by the light of bonfires,— a practice which they prefer to mere relaxation or sleep, on their days of rest.

The bark in which they passed the Gulf of Cariaco was commodious, and large skins of the jaguar were spread for their repose during the night. The cold, however, prevented them from sleeping, although, as they were surprised to find, the thermometer was as high as 71·2°. The circumstance that in a warm country a degree of cold which would be productive of no inconvenience to the inhabitant of a temperate climate, excites a disagreeable feeling, is worthy of the attention of physiologists. When Bouguer reached the summit of Pelée, in the island of Martinico, he trembled with cold, although the heat was above 70·7°; and in heavy showers at Cumana, when the thermometer indicates the same temperature, the inhabitants make bitter complaints.

About eight in the morning they landed at the point of Araya, near the new salt-works, which are situated in a plain destitute of vegetation. From this spot are seen the islet of Cubagua, the lofty hills of Margaretta, the ruins of the castle of St Jago, the Cerro de la Vela, and the limestone ridge of the Bergantin, bounding the horizon toward the south. Here salt is procured by digging brine-pits in the clayey soil, which is impregnated with muriate of soda. In 1799 and 1800, the consumption of this article in the provinces of Cumana and Barcelona amounted to 9000 or 10,000 *fanegas*, each 16 *arrobas*, or 405¾ lbs. avoirdupois. Of this quantity the salt-works of Araya yield only about a third part; the rest being obtained from sea-water in the Morro of Barcelona, at Pozuelos, at Piritu, and in the Golfo Triste.

In order to understand the geological relations of this saliferous clay, it is necessary to follow our author in his exposition of the nature of the neighbouring country. Three great parallel chains of mountains extend from east to west. The two most northerly, which are primitive, constitute the cordilleras of the island of Margaretta, as well as of Araya. The most southerly, the cordillera of Bergantin and Cocollar, is secondary, although more elevated than the others. The two former have been separated by the sea, and the islets of Coche and Cubagua are supposed to be remnants of the submersed land. The Gulf of Cariaco divides the chains of Araya and Cocollar, which were connected, to the east of the town of Cariaco, between the lakes of Campoma and Putaquao, by a kind of dike. This barrier, which had the name of Cerro de Meapire, prevented, in remote times, the waters of the Gulf of Cariaco from uniting with those of the Gulf of Paria.

The western slope of the peninsula of Araya, and the plains on which rises the castle of St Antony, are covered with recent deposites of sandstone, clay, and gypsum. Near Manifuarez, a conglomerate with calcareous cement rests on the mica-slate; while on the opposite side, near Punta Delgada, it is superimposed on a compact bluish-gray limestone, containing a few organic remains, traversed by small veins of calcareous spar, and analogous to that of the Alps.

The saliferous clay is generally of a smoke-gray colour, earthy and friable, but encloses masses of a dark-brown tint and more solid texture. Selenite and fibrous gypsum are disseminated in it. Scarcely any shells are to be seen, although the adjacent rocks contain abundance of them. The muriate of soda is not discoverable by the naked eye; but when a mass is sprinkled with rain-water and exposed to the sun, it appears in large crystals. In the marsh to the east of the castle of St Jago, which receives only rain-water, crystallized and very pure muriate of soda forms, after great droughts, in masses of large size. The new salt-works of Araya have five very extensive reservoirs, with a depth of eight inches, and are supplied partly with sea-water and partly with rain. The evaporation is so rapid, that salt is collected in eighteen or twenty days after they are filled; and it is freer from earthy muriates and sulphates than that of Europe, although manufactured with less care.

After examining these works, they departed at the decline of day, and proceeded toward an Indian cabin some miles distant. Night overtook them in a narrow path between a range of perpendicular rocks and the sea. Arriving at the foot of the old castle of Araya, which stands on a bare and arid mountain, and is crowned with agave, columnar cactus, and prickly mimosas, they were desirous of stopping to admire the majestic spectacle, and observe the setting of the planet Venus; but their guide, who was parched with thirst, earnestly urged them to return, and hoped to work on their fears by continually warning them of jaguars and rattlesnakes. They at length yielded to his solicitations; but, after proceeding three-quarters of an hour along a shore covered by the tide, they were joined by the negro that carried their provisions, who led them through a wood of nopals to the hut of an Indian, where they were received with cordial hospitality. The several classes of natives in this district live by catching fish, part of which they carry to Cumana. The wealth of the inhabitants consists chiefly of goats, which are of a very large size, and brownish-yellow colour. They are marked like the mules, and roam at large.

E

Among the mulattoes, whose hovels surrounded the salt-lake near which they had passed the night, they found an indigent Spanish cobbler, who received them with an air of gravity and importance. After amusing them with a display of his knowledge, he drew from a leathern bag a few very small pearls, which he forced them to accept, enjoining them to note on their tablets, " that a poor shoemaker of Araya, but a white man, and of noble Castilian descent, was enabled to give them what, on the other side of the sea, would be sought for as a thing of great value."

The pearl-shell (*Avicula margaritifera*) is abundant on the shoals which extend from Cape Paria to the Cape of Vela. Margarita, Cubagua, Coche, Punta Araya, and the mouth of the Rio la Hacha, were as celebrated in the sixteenth century for them, as the Persian Gulf was among the ancients. At the beginning of the conquest, the island of Coche alone furnished 1500 marks (1029 Troy pounds) monthly. The portion which the king's officers drew from the produce of the pearls amounted to £3406, 5s.; and it would appear, that up to 1530, the value of those sent to Europe amounted, at a yearly average, to more than £130,000. Towards the end of the sixteenth century, this fishery diminished rapidly; and, according to Laet, had been long given up in 1683. The artificial imitations, and the great diminution of the shells, rendered it less lucrative. At present, the Gulf of Panama and the mouth of the Rio de la Hacha are the only parts of South America in which this branch of industry is continued.

On the morning of the 20th, a young Indian conducted the travellers over Barigon and Caney, to the village of Maniquarez. The thermometer kept as high as 78·5°, and before their guide had travelled a league, he frequently sat down to rest himself, and expressed a desire to repose under the shade of a tamarind-tree until night should approach. Humboldt explains the circumstance, that the natives complain more of lassitude under an intense heat than Europeans not inured to it, by a reference to their listless disposition, and their not being excited by the same stimulus.

In crossing the arid hills of Cape Cirial, they perceived a strong smell of petroleum, the wind blowing from the side where the springs of that substance occur. Near the village of Maniquarez, they found the mica-slate cropping out from below the secondary rocks. It was of a silvery white, contained garnets, and was traversed by small layers of quartz. From a detached block of this last, found on the shore, they separated a fragment of cyanite, the only specimen of that mineral seen by them in South America.

A rude manufacture of pottery is carried on at that hamlet by the Indian women. The clay is produced by the decomposition of mica-slate, and is of a reddish colour. The natives, being unacquainted with the use of ovens, place twigs around the vessels, and bake them in the open air.

At the same place they met with some Creoles who had been hunting small deer in the uninhabited islet of Cubagua, where they are very abundant. These creatures are of a brownish-red hue, spotted with white, and of the latter colour beneath. They belong to the species named by naturalists *Cervus Mexicanus*.

In the estimation of the natives, the most curious production of the coast of Araya is what they call the *eye-stone*. They consider it as both a stone and an animal, and assert, that when it is found in the sand it is motionless; whereas on a polished surface, as an earthen plate, it moves when stimulated by lemon-juice. When introduced into the eye, it expels every other substance that may have accidentally insinuated itself. The people offered these stones to the travellers by hundreds, and wished to put sand into their eyes, that they might try the power of this wondrous remedy; which, however, was nothing else than the operculum of a small shell-fish.

Near Cape de la Brea, at the distance of eighty feet from the shore, is a small stream of naphtha, the produce of which covers the sea to a great extent. It is a singular circumstance that this spring issues from mica-slate, all others that are known belonging to secondary deposites.

After examining the neighbourhood of Maniquarez, the adventurers embarked at night in a small fishing-boat, so leaky that a person was constantly employed in baling out the water with a calabash, and arrived in safety at Cumana.

CHAPTER VII.

Missions of the Chaymas.

Excursion to the Missions of the Chayma Indians—Remarks on Cultivation—The Impossible—Aspect of the Vegetation—San Fernando—Account of a Man who suckled a Child—Cumanacoa —Cultivation of Tobacco—Igneous Exhalations—Jaguars— Mountain of Cocollar—Turimiquiri—Missions of San Antonio and Guanaguana.

ON the 4th of September, at an early hour, our travellers commenced an excursion to the missionary stations of the Chayma Indians, and to the lofty mountains which traverse New Andalusia. The morning was deliciously cool; and from the summit of the hill of San Francisco they enjoyed in the short twilight an extensive view of the sea, the adjacent plain, and the distant peaks. After walking two hours they arrived at the foot of the chain, where they found different rocks, together with a new and more luxuriant vegetation. They observed that the latter was more brilliant wherever the limestone was covered by a quartzy sandstone,—a circumstance which probably depends not so much on the nature of the soil as on its greater humidity; the thin layers of slate-clay which the latter contains preventing the water from filtering into the crevices of the former. In those moist places they always discovered appearances of cultivation, huts inhabited by mestizoes, and placed in the centre of small enclosures, containing papaws, plantains, sugar-canes, and maize. In Europe, the wheat, barley, and other kinds of grain, cover vast spaces of ground, and, in general, wherever the inhabitants live upon corn, the cultivated lands are not separated from each other by the intervention of large wastes; but in the torrid zone, where the fertility of the soil is proportionate to the heat and humidity of the air, and where man has appropriated plants that yield earlier and more abundant crops, an immense population finds ample subsistence on a narrow space. The scattered disposition of the huts in the midst of the forest indicates to the traveller the fecundity of nature.

In so mild and uniform a climate, the only urgent want of man is that of food; and in the midst of abundance, his intellectual faculties receive less improvement than in colder regions, where his necessities are numerous and diversified. While in Europe, we judge of the inhabitants of a country by the extent of laboured ground; in the warmest parts of South America, populous provinces seem to the traveller almost deserted, because a very small extent of soil is sufficient for the maintenance of a family. The insulated state in which the natives thus live, prevents any rapid progress of civilisation, although it develops the sentiments of independence and liberty.

As the travellers penetrated into the forest, the barometer indicated the progressive elevation of the land. About three in the afternoon they halted on a small flat, where a few houses had been erected near a spring, the water of which they found delicious. Its temperature was 72·5°, while that of the air was 83·7°. From the top of a sandstone-hill in the vicinity, they had a splendid view of the sea and part of the coast, while in the intervening space, the tops of the trees, intermixed with flowery lianas, formed a vast carpet of deep verdure. As they advanced toward the south-west, the soil became dry and loose. They ascended a group of rather high mountains, destitute of vegetation, and having steep declivities. This ridge is named the *Impossible,* it being imagined that, in case of invasion, it might afford a safe retreat to the inhabitants of Cumana. The prospect was finer and more extensive than from the fountain above mentioned.

They arrived on the summit only a little before dusk. The setting of the sun was accompanied by a very rapid diminution of temperature, the thermometer suddenly falling from 77·4° to 70·3°, although the air was calm. They passed the night in a house at which there was a military post of eight men, commanded by a Spanish sergeant. When, after the capture of Trinidad by the English in 1797, Cumana was threatened, many of the people fled to Cumanacoa, leaving the more valuable of their property in sheds constructed on this ridge. The solitude of the place reminded Humboldt of the nights which he had passed on the top of St Gothard. Several parts of the surrounding forests were burning, and the reddish flames arising amidst clouds of smoke, presented a most impressive spectacle. The shepherds set fire to the woods for the purpose of improving the pasturage, though conflagrations are often caused by the negligence of the wandering Indians. The number of old trees on the road from Cumana to Cumanacoa has been greatly reduced by these accidents; and in several parts of the province the dryness has increased, owing both to the diminution of the forests, and the frequency of earthquakes which produce crevices in the soil.

Leaving the Impossible on the 5th before sunrise, they descended by a very narrow path bordering on precipices. The summit of the ridge was of quartzy sandstone, beneath which the Alpine limestone reappeared. The strata being generally inclined to the south, numerous springs gush out on that side, and in the rainy season form torrents which fall in cascades, shaded by the hura, the cuspa, and the trumpet-tree. The cuspa, which is common in the neighbourhood of Cumana, had long been used for carpenter-work, but has of late attracted notice as a powerful tonic or febrifuge.

Emerging from the ravine which opens at the foot of the mountain, they entered a dense forest, traversed by numerous small rivers, which were easily forded. They observed that the leaves of the cecropia were more or less silvery according as the soil was dry or marshy, and specimens occurred in which they were entirely green on both sides. The roots of these shrubs were concealed beneath tufts of dorstenia, a plant which thrives only in shady and moist places. In the midst of the forest they found papaws and orange-trees bearing excellent fruit, which they conjectured to be the remains of some Indian plantations, as in these countries they are no more indigenous than the banana, the maize, the manioc, and the many other useful plants whose native country is unknown, although they have accompanied man in his migrations from the most remote periods.

" When a traveller newly arrived from Europe,"

VEGETATION OF NEW ANDALUSIA. 89

says Humboldt, " penetrates for the first time into the forests of South America, nature presents herself to his view in an unexpected aspect; the objects by which he is surrounded bear but a faint resemblance to the pictures drawn by celebrated writers on the banks of the Mississippi, in Florida, and in other temperate regions of the New World. He perceives at every step, that he is not upon the verge, but in the centre of the torrid zone,—not in one of the West India islands, but upon a vast continent, where the mountains, the rivers, the mass of vegetation, and every thing else, are gigantic. If he be sensible to the beauties of rural scenery, he finds it difficult to account to himself for the diversified feelings which he experiences: he is unable to determine what most excites his admiration; whether the solemn silence of the wilderness, or the individual beauty and contrast of the forms, or the vigour and freshness of vegetable life that characterize the climate of the tropics. It might be said that the earth, overloaded with plants, does not leave them room·enough for growth. The trunks of the trees are every where covered with a thick carpet of verdure; and were the orchideæ and the plants of the genera Piper and Pothos, which grow upon a single courbaril or American fig-tree, transferred to the ground, they would cover a large space. By this singular denseness of vegetation, the forests, like the rocks and mountains, enlarge the domain of organic nature. The same lianas, which creep along the ground, rise to the tops of the trees, and pass from the one to the other at a height of more than a hundred feet. In consequence of this intermixture of parasitic plants, the botanist is often led to confound the flowers, fruits, and foliage, which belong to different species."

The philosophers walked for some hours under the shade of these arches, which scarcely admitted an occasional glimpse of the clear blue sky, and for the first time admired the pendulous nests of the orioles, which mingled their warblings with the cries of the parrots and macaws. The latter fly only in pairs, while the former are seen in flocks of several hundreds. At the distance of about a league from the village of San Fernando, they issued from the woods, and entered an open country, covered with aquatic plants from eight to ten feet high; there being no meadows or pastures in the lower parts of the torrid zone as in Europe. The road was bordered with a kind of bamboo rising more than forty feet. These plants, according to Humboldt, are less common in America than is usually supposed, although they form dense woods in New Grenada and Quito, and occur abundantly on the western slope of the Andes.

They now entered San Fernando, which is situated in a narrow plain, and bounded by limestone rocks. This was the first missionary station they saw in America. The houses of the Chayma Indians were built of clay, strengthened by lianas, and the streets were straight, and intersected each other at right angles. The great square in the centre of the village contains the church, the house of the missionary, and another, destined for the accommodation of travellers, which bears the pompous name of the king's house (Casa del Rey). These royal residences occur in all the Spanish settlements, and are of the greatest benefit in countries where there are no inns.

They had been recommended to the friars, who

superintend the missions of the Chaymas, by their syndic at Cumana, and the superior, a corpulent and jolly old capuchin, received them with kindness. This respectable personage, seated the greater part of the day in an arm-chair, complained bitterly of the indolence of his countrymen. He considered the pursuits of the travellers as useless, smiled at the sight of their instruments and dried plants, and maintained that of all the enjoyments of life, without excepting sleep, none could be compared with the pleasure of eating good beef.

This mission was founded about the end of the seventeenth century, near the junction of the Manzanares and Lucasperez; but, in consequence of a fire, was removed to its present situation. The number of families now amounted to a hundred, and, as the head of the establishment observed, the custom of marrying at a very early age contributes greatly to the rapid increase of population.

In the village of Arenas, which is inhabited by Indians of the same race as those of San Fernando, there lived a labourer, Francisco Lozano, who had suckled a child. Its mother happening to be sick, he took it, and in order to quiet it, pressed it to his breast, when the stimulus imparted by the sucking of the child caused a flow of milk. The travellers saw the certificate drawn up on the spot to attest this remarkable fact, of which several eyewitnesses were still living. The man was not at Arenas during their stay at the mission, but afterwards visited them at Cumana, accompanied by his son, when M. Bonpland examined his breasts, and found them wrinkled, like those of women who have nursed. He was not an Indian, but a white descended from

European parents. Alexander Benedictus relates a similar case of an inhabitant of Syria, and other authors have given examples of the same nature.

Returning towards Cumana, they entered the small town of Cumanacoa, situated in a naked and almost circular plain, surrounded by lofty mountains, and containing about two thousand three hundred inhabitants. The houses were low and slight, and with very few exceptions built of wood. The travellers were surprised to find the column of mercury in the barometer scarcely 7·3 lines shorter than on the coast. The hollow in which the town is erected is not more than 665 feet above the level of the sea, and only seven leagues from Cumana; but the climate is much colder than in the latter place, where it scarcely ever rains; whereas at Cumanacoa there are seven months of severe weather. It was during the winter season that our travellers visited the missions. A dense fog covered the sky every night; the thermometer varied from 64·8° to 68°; and Deluc's hygrometer indicated 85°. At ten in the morning the thermometer did not rise above 69·8°, but from noon to three o'clock attained the height of from 78·8° to 80·6°. About two, large black clouds regularly formed, and poured down torrents of rain, accompanied by thunder. At five the rain ceased, and the sun reappeared; but at eight or nine the fog again commenced. In consequence of the humidity, the vegetation, although not very diversified, is remarkable for its freshness. The soil is highly fertile; but the most valuable production of the district is tobacco, the cultivation of which in the province of Cumana is nearly confined to this valley.

Next to the tobacco of Cuba and the Rio Negro,

that grown here is the most aromatic. The seed is sown in the beginning of September, and the cotyledons appear on the eighth day. The young plants are then covered with large leaves to protect them from the sun. A month or two after, they are transferred to a rich and well-prepared soil, and disposed in rows, three or four feet distant from each other. The whole is carefully weeded, and the principal stalk is several times topped, until the leaves are mature, when they are gathered. They are then suspended by threads of the *Agave Americana*, and their ribs taken out; after which they are twisted. The cultivation of tobacco was a royal monopoly, and employed about 1500 persons. Indigo is also raised in the valley of Cumanacoa.

This singular plain appeared to be the bed of an ancient lake. The surrounding mountains are all precipitous, and the soil contains pebbles and bivalve shells. One of the gaps in the range, they were informed, was inhabited by jaguars, which passed the day in caves, and roamed about the plantations at night. The preceding year, one of them had devoured a horse belonging to a farm in the neighbourhood. The groans of the dying animal awoke the slaves, who went out armed with lances and large knives, with which they despatched the tiger after a vigorous resistance.

From two caverns in this ravine there at times issue flames, which illumine the adjacent mountains, and are seen to a great distance at night. The phenomenon was accompanied by a long-continued subterraneous noise at the time of the last earthquake. A first attempt to penetrate into this pass was rendered unsuccessful, by the strength of the vegetation and the intertwining of lianas and thorny plants; but the inhabitants becoming interested in the researches of the travellers, and being desirous to know what the German miner thought of the gold ore which they imagined to exist in it, cleared a path through the woods. On entering the ravine they found traces of jaguars; and the Indians returned for some small dogs, upon which they knew these animals would spring in preference to attacking a man. The rocks that bound it are perpendicular, and what geologists term Alpine limestone. The excursion was rendered hazardous by the nature of the ground; but they at length reached the pretended gold mine, which was merely an excavation in a bed of black marl containing iron pyrites, a substance which the guides insisted was no other than the precious metal.

They continued to penetrate into the crevice, and after undergoing great fatigue, reached a wall of rock, which, rising perpendicularly to the height of 5116 feet, presented two inaccessible caverns inhabited by nocturnal birds. Halting at the foot of one of the caves from which flames had been seen to issue, they listened to the remarks of the natives respecting the probability of an increase in the frequency of the agitations to which New Andalusia had so often been subjected. The cause of the luminous exhalations, however, they were unable to ascertain.

On the 12th they continued their journey to the convent of Caripe, the principal station of the Chayma missions, choosing, instead of the direct road, the line of the mountains Cocollar and Turimiquiri. At the Hato de Cocollar, a solitary farm situated on a small elevated plain, they rested for some time, and had the good fortune to enjoy at once a delightful climate and the hospitality of the proprietor.

From this elevated point, as far as the eye could reach, they saw only naked savannahs, although in the neighbouring valleys they found tufts of scattered trees, and a profusion of beautiful flowers. The upper part of the mountain was destitute of wood, though covered with gramineous plants,—a circumstance which Humboldt attributes more to the custom of burning the forests than to the elevation of the ground, which is not sufficient to prevent the growth of trees.

Their host, Don Mathias Yturburi, a native of Biscay, had visited the New World with an expedition, the object of which was to form establishments for procuring timber for the Spanish navy. But these natives of a colder climate were unable to support the fatigue of so laborious an occupation, the heat, and the effect of noxious vapours. Destructive fevers carried off most of the party, when this individual withdrew from the coast, and settling on the Cocollar, became the undisturbed possessor of five leagues of savannahs, among which he enjoyed independence and health.

" Nothing," says Humboldt, " can be compared to the impression of the majestic tranquillity left on the mind by the view of the firmament in this solitary place. Following with the eye, at eveningtide, those meadows which stretch along the horizon, and the gently-undulated plain covered with plants, we thought we saw in the distance, as in the deserts of the Orinoco, the surface of the ocean supporting the starry vault of heaven. The tree under which we were seated, the luminous insects that vaulted in the air, and the constellations which shone in the south, seemed to tell us that we were far from our native land. In the midst of this exotic nature, when the bell of a cow or the lowing of a bull was heard from the bottom of a valley, the remembrance of our country was suddenly awakened by the sounds. They were like distant voices, that came from beyond the ocean, and by the magic of which we were transported from the one hemisphere to the other. Strange mobility of the human imagination, the never-failing source of our enjoyments and griefs!"

In the cool of the morning, they commenced the ascent of Turimiquiri, the summit of the Cocollar, which with the Brigantine forms a mass of mountains, formerly named by the natives the Sierra de los Tageres. They travelled part of the way on horses, which are left to roam at large in these wilds, though some of them have been trained to the saddle. Stopping at a spring which issued from a bed of quartzy sandstone, they found its temperature to be 69·8°. To the height of 4476 feet, this mountain, like those in its vicinity, was covered with gramineous plants. The pastures became less rich in proportion to the elevation, and wherever the scattered rocks afforded a shade lichens and mosses occurred. The summit is 4521 feet above the level of the sea. The view from it was extensive and highly picturesque: chains of mountains, running from east to west, enclosed longitudinal valleys, which were intersected at right angles by numberless ravines. The distant peninsula of Araya formed a dark streak on a glittering sea, and the more distant rocks of Cape Macanao rose amidst the waters like an immense rampart.

On the 14th of September, they descended the Cocollar in the direction of San Antonio, where was also a mission. After passing over savannahs strewed with blocks of limestone, succeeded by a dense forest and two very steep ridges, they came to a beautiful valley, about twenty miles in length, in which are situated the missions of San Antonio and Guanaguana. Stopping at the former only to open the barometer and take a few altitudes of the sun, they forded the rivers Colorado and Guarapiche, and proceeding along a level and narrow road covered with thick mud, amid torrents of rain, reached in the evening the latter of these stations, where they were cordially received by the missionary. This village had existed only thirty years on the spot which it then occupied, having been transferred from a place more to the south. Humboldt remarks, that the facility with which the Indians remove their dwellings is astonishing, there being several small towns in South America which have thrice changed their situation in less than half a century. These compulsory migrations are not unfrequently caused by the caprice of an ecclesiastic; and as the houses are constructed of clay, reeds, and palm-leaves, a hamlet shifts its position like a camp.

The mission of San Antonio had a small church with two towers, built of brick and ornamented with Doric columns, the wonder of the country; but that of Guanaguana possessed as yet no place of worship, although a spacious house had been built for the padre, the terraced roof of which was ornamented with numerous chimneys like turrets, and which, he informed the travellers, had been erected for no other purpose than to remind him of his native country. The Indians cultivate cotton. The machines by which they separate the wool from the seeds are of very simple construction, consisting of wooden cylinders of very small diameter, made to revolve by a treadle. Maize is the article on which they

F

principally depend for food; and when it happens to be destroyed by a protracted drought, they betake themselves to the surrounding forests, where they find subsistence in succulent plants, cabbage-palms, fern-roots, and the produce of various trees.

Proceeding towards the valley of Caripe, the travellers passed a limestone ridge which separates it from that of Guanaguana,—an undertaking which they found rather difficult, the path being in several parts only fourteen or fifteen inches broad, and the slopes being covered with very slippery turf. When they had reached the summit, an interesting spectacle presented itself to their view, consisting of the vast savannahs of Maturin and Rio Tigre, the Peak of Turimiquiri, and a multitude of parallel hills resembling the waves of a troubled ocean.

Descending the height by a winding path, they entered a woody country, where the ground was covered by moss and a species of *Drosera*. As they approached the convent of Caripe, the forests grew more dense, and the power of vegetation increased. The calcareous strata became thinner, forming graduated terraces, while the stone itself assumed a white colour, with a smooth or imperfectly conchoidal fracture. This rock Humboldt considers as analogous to the Jura deposites. He found the level of the valley of Caripe 1279 feet higher than that of Guanaguana. Although the former is only separated from the latter by a narrow ridge, it affords a complete contrast to it, being deliciously cool and salubrious, while the other is remarkable for its great heat.

CHAPTER VIII.

Excursion continued, and Return to Cumana.

Convent of Caripe—Cave of Guacharo, inhabited by Nocturnal Birds—Purgatory—Forest Scenery—Howling Monkeys—Vera Cruz—Cariaco — Intermittent Fevers — Cocoa-trees — Passage across the Gulf of Cariaco to Cumana.

ARRIVING at the hospital of the Arragonese Capuchins, which was backed by an enormous wall of rocks of resplendent whiteness, covered with a luxuriant vegetation, our travellers were hospitably received by the monks. The superior was absent; but having heard of their intention to visit the place, he had provided for them whatever could serve to render their abode agreeable. The inner court, surrounded by a portico, they found highly convenient for setting up their instruments and making observations. In the convent they found a numerous society, consisting of old and infirm missionaries, who sought for health in the salubrious air of the mountains of Caripe, and younger ones newly arrived from Spain. Although the inmates of this establishment knew that Humboldt was a Protestant, they manifested no mark of distrust, nor proposed any indiscreet question, to diminish the value of the benevolence which they exercised with so much liberality. Even the light of science had in some degree extended to this obscure place; for, in the library of the superior, they found among other books the Traité d'Electricité by the Abbé Nollet, and one of the monks had brought with him a Spanish translation of Chaptal's Treatise on Chemistry.

The height of this monastery above the sea is nearly the same as that of Caraccas and the inhabited parts of the Blue Mountains of Jamaica. The thermometer was between 60·8° and 63° at midnight, between 66·2° and 68° in the morning, and only 69·8° or 72·5° about one o'clock. The mean temperature, inferred from that of the month of September, appears to be 65·3°. This degree of heat is sufficient to develop the productions of the torrid zone, although much inferior to that of the plains of Cumana. Water exposed in vessels of porous clay cools during the night as low as 55·4°. The mild climate and rarefied air of this place have been found highly favourable to the cultivation of coffee, which was introduced into the province by the prefect of the Capuchins, an active and enlightened man. In the garden of the community were many culinary vegetables, maize, the sugar-cane, and five thousand coffee-trees.

The greatest curiosity in this beautiful and salubrious district is a cavern inhabited by nocturnal birds, the fat of which is employed in the missions for dressing food. It is named the Cave of Guacharo, and is situated in a valley three leagues distant from the convent.

On the 18th of September our travellers, accompanied by most of the monks and some of the Indians, set out for this aviary, following for an hour and a half a narrow path, leading across a fine plain covered with beautiful turf; then, turning westward along a small river which issues from the cave, they proceeded, during three quarters of an hour, sometimes walking in the water, sometimes on a slippery and miry soil between the torrent and a wall of rocks, until they arrived at the foot of the lofty mountain of Guacharo. Here the torrent ran in a deep ravine, and they went on under a projecting cliff which prevented them from seeing the sky, until at the last turning they came suddenly upon the immense opening of the recess, which is eighty-five feet broad and seventy-seven feet high. The entrance is toward the south, and is formed in the vertical face of a rock, covered with trees of gigantic height, intermixed with numerous species of singular and beautiful plants, some of which hang in festoons over the vault. This luxuriant vegetation is not confined to the exterior of the cave, but appears even in the vestibule, where the travellers were astonished to see heliconias nineteen feet in height, palms, and arborescent arums. They had advanced about four hundred and sixty feet before it became necessary to light their torches, when they heard from afar the hoarse screams of the birds.

The guacharo is the size of a domestic fowl, and has somewhat the appearance of a vulture, with a mouth like that of a goatsucker. It forms a distinct genus in the order *Passeres*, differing from that just named in having a stronger beak, furnished with two denticulations, though in its manners it bears an affinity to it as well as to the Alpine crow. Its plumage is dark bluish-gray, minutely streaked and spotted with deep-brown, the head, wings, and tail, being marked with white spots bordered with black. The extent of the wings is three feet and a half. It lives on fruits, but quits the cave only in the evening. The shrill and piercing cries of these birds, assembled in multitudes, are said to form a harsh and disagreeable noise, somewhat resembling that of a rookery. The nests, which the guides showed by means of torches fastened to a long pole, were placed in funnel-shaped holes in the roof. The noise increased as they advanced, the animals being frightened by the numerous lights.

About midsummer every year, the Indians armed with poles enter the cave, and destroy the greater part of the nests. Several thousands of young birds are thus killed, and the old ones hover around, uttering frightful cries. Those which are secured in this manner are opened on the spot, to obtain the fat which exists abundantly in their abdomen, and which is subsequently melted in clay vessels, over fires of brushwood. This substance is semifluid, transparent, destitute of smell, and keeps above a year without becoming rancid. At the convent of Caripe it was used in the kitchen of the monks, and our travellers never found that it communicated any disagreeable smell or taste to the food.

The guacharoes would have been long ago destroyed, had not the superstitious dread of the Indians prevented them from penetrating far into the cavern. It also appears, that birds of the same species dwell in other inaccessible places in the neighbourhood, and that the great cave is repeopled by colonies from them. The hard and dry fruits which are found in the crops and gizzards of the young ones are considered as an excellent remedy against intermittent fevers, and regularly sent to Cariaco and other parts of the lower districts where such diseases prevail.

The travellers followed the banks of the small river which issues from the cavern as far as the mounds of calcareous incrustations permitted them, and afterwards descended into its bed. The cave preserved the same direction, breadth, and height, as at its entrance, to the distance of 1554 feet. The natives having a belief that the souls of their ancestors inhabit its deep recesses, the Indians who accompanied our travellers could hardly be persuaded to venture into it. Shooting at random in the dark, they obtained two specimens of the guacharo. Having proceeded to a certain distance, they came to a mass of stalactite, beyond which the cave became narrower, although it retained its original direction. Here the rivulet had deposited a blackish mould resembling that observed at Muggendorf in Franconia. The seeds, which the birds carry to their young, spring up wherever they are dropped into it; and M. Humboldt and his friend were astonished to find blanched stalks that had attained a height of two feet.

As the missionaries were unable to persuade the Indians to advance farther, the party returned. The river, sparkling amid the foliage of the trees, seemed like a distant picture, to which the mouth of the cave formed a frame. Having sat down at the entrance to enjoy a little needful repose, they partook of a repast which the missionaries had prepared, and in due time returned to the convent.

The days which our travellers passed at this religious house glided hastily and pleasantly past. From morning to night they traversed the forests and mountains collecting plants; and when the rains prevented them from making distant excursions; they visited the huts of the Indians; returning to the good monks only when the sound of the bell called them to the solace of the refectory. Sometimes also they followed them to the church, to witness the religious instruction given to the Indians; which was found a difficult task, owing to the imperfect knowledge of the Spanish language possessed by the latter. The evenings were employed in taking notes, drying plants, and sketching those that appeared new.

The natural beauties of this interesting valley engaged them so much, that they were long in perceiving the embarrassment felt by their kind entertainers, who had now but a very slender store of wine and bread. At length, on the 22d September, they departed, followed by four mules carrying their instruments and plants. The descent of the rugged chain of the Brigantine and Cocollar, which is about 4400 feet in height, is exceedingly difficult. The missionaries have given the name of Purgatory to an extremely steep and slippery declivity at the base of a sandstone rock, in passing which the mules, drawing their hind legs under their bodies, slide down at a venture. From this point they saw toward the left the great peak of Guacharo, which presented a very picturesque appearance; and soon after entered a dense forest, through which they descended for seven hours in a kind of ravine, the path being formed of steps from two to three feet high, over which the mules leaped like wild goats. The creoles have sufficient confidence in these animals to remain in their saddles during this dangerous passage; but our travellers preferred walking.

The forest was exceedingly dense, and consisted

of trees of stupendous size. The guides pointed out some whose height exceeded 130 feet, while the diameter of many of the curucays and hymendas was more than three yards. Next to these, the plants which most attracted their notice were the dragon's-blood (*Croton sanguifluum*), the purple juice of which flowed along the whitish bark, various species of palms, and arborescent ferns of large size. The old trunks of some of the latter were covered with a carbonaceous powder, having a metallic lustre like graphite.

As they descended the mountain the tree-ferns diminished, while the number of palms increased. Large-winged butterflies (*nymphales*) became more common, and every thing showed that they were approaching the coast. The weather was cloudy, the heat oppressive, and the howling of the monkeys gave indication of a coming thunder-storm. These creatures, the arguatoes, resemble a young bear, and are about three feet long from the top of the head to the root of the tail. The fur is tufty and reddish-brown, the face blackish-blue, with a bare and wrinkled skin, and the tail long and prehensile.

While engaged in observing a troop of them cross the road upon the horizontal branches of the trees, the travellers met a company of naked Indians proceeding towards the mountains of Caripe. The men were armed with bows and arrows, and the women, heavily laden, brought up the rear. They marched in silence, with their eyes fixed on the ground. Our philosophers, oppressed with the increasing heat and faint with fatigue, endeavoured to learn from them the distance of the missionary convent of Vera Cruz, where they intended to pass the night; but little information could be obtained on account of their imperfect knowledge of the Spanish language.

Continuing to descend amid scattered blocks, they unexpectedly found themselves at the end of the forest, when they entered a savannah, the verdure of which had been renewed by the winter rains. Here they had a splendid view of the Sierra del Guacharo, the northern declivity of which presented an almost perpendicular wall, exceeding 3200 feet in height, and scantily covered with vegetation. The ground before them consisted of several level spaces, lying above each other like vast steps. The mission of Vera Cruz, which is situated in the middle of it, they reached in the evening, and next day continued their journey toward the Gulf of Cariaco.

Proceeding on their way, they entered another forest, and reached the station of Catuaro, situated in a very wild spot, where they lodged at the house of the priest. Their host was a doctor of divinity, a thin little man, of petulant vivacity, who talked continually of a lawsuit in which he was engaged with the superior of his convent, and wished to know what Humboldt thought of free-will and the souls of animals. At this place they met with the corregidor of the district, an amiable person, who gave them three Indians to assist in cutting a way through the forest, the lianas and intertwining branches having obstructed the narrow lanes. The little missionary, however, insisted on accompanying them to Cariaco, and contrived to render the road extremely tedious by his observations on the necessity of the slave-trade, the innate wickedness of blacks, and the benefit which they derived from being reduced to bondage by Christians.

The road which they followed through the forest of Catuaro resembled that of the preceding day. The clay, which filled the path and rendered it excessively slippery, was produced by layers of sandstone and slate-clay which cross the calcareous strata. At length, after a fatiguing march, they reached the town of Cariaco, on the coast, where they found a great part of the inhabitants confined to their beds with intermittent fever. The low situation of the place as well as of the surrounding district, the great heat and moisture, and the stagnant marshes generated during the rainy season, are supposed to be the causes of this disease, which often assumes a malignant character, and is accompanied with dysentery. Men of colour, and especially creole negroes, resist the influence of the climate much better than any other race. It is generally observed, however, that the mortality is less than might be supposed; for although intermittent fevers, when they attack the same individual several years in succession, alter and weaken the constitution, they do not usually cause death. It is remarkable, that the natives believe the air to have become more vitiated in proportion as a larger extent of land has been cultivated; but the miasmata from the marshes, and the exhalations from the mangroves, avicenniæ, and other astringent plants growing on the borders of the sea, are probably the real causes of the unhealthiness of the coasts.

In 1800 the town of Cariaco contained more than 6000 inhabitants, who were actively employed in the cultivation of cotton, the produce of which exceeded 10,000 quintals (9057 ℔s. avoirdupois). The capsules, after the separation of the wool, were carefully burnt, as they were thought to occasion noxious exhalations when thrown into the river. Cacao and sugar were also raised to a considerable extent.

As our travellers were not sufficiently inured to the climate, they considered it prudent to leave Cariaco as expeditiously as possible on account of the fever. Embarking early in the morning, they proceeded westward along the river of Carenicuar, which flows through a deep marshy soil covered with gardens and plantations of cotton. The Indian women were washing their linen with the fruit of the parapara (*Sapindus saponaria*). Contrary winds, accompanied with heavy rain and thunder, rendered the voyage disagreeable; more especially as the canoe was narrow and overloaded with raw sugar, plantains, cocoa-nuts, and passengers. Swarms of flamingoes, egrets, and cormorants, were flying toward the shore, while the alcatras, a large species of pelican, less affected by the weather, continued fishing in the bay. The general depth of the sea is from 288 to 320 feet; but at the eastern extremity of the gulf it is only from nineteen to twenty-five feet for an extent of seventeen miles, and there is a sandbank, which at low water resembles a small island. They crossed the part where the hot springs rush from the bottom of the ocean; but it being high water the change of temperature was not very perceptible. The contrary winds continuing, they were forced to land at Pericautral, a small farm on the south side of the gulf. The coast, although covered by a beautiful vegetation, was almost destitute of human labour, and scarcely possessed seven hundred inhabitants. The cocoa-tree is the principal object of cultivation. This palm thrives best in the neighbourhood of the sea, and like the sugar-cane, the plantain, the mammee-apple, and the alligator-pear, may be watered either with fresh or salt water. In other parts of America it is generally nourished around farm-houses; but along the Gulf of Cariaco it forms real plantations, and at Cumana they talk of a *hacienda de coco*, as they do of a *hacienda de canna*, or *de cacao*. In moist and fertile ground it begins to bear abundantly the fourth year; but in dry soils it does not produce fruit until the tenth. Its duration does not generally exceed ninety or a hundred years; at which period its mean height is about eighty feet. Throughout this coast a cocoa-tree supplies annually about a hundred nuts, which yield eight flascos of oil. The flasco is sold for about sixteenpence. A great quantity is made at Cumana, and Humboldt frequently witnessed the arrival there of canoes containing 3000 nuts. The oil, which is clear and destitute of smell, is well adapted for burning.

After sunset they left the farm of Pericautral, and at three in the morning reached the mouth of the Manzanares, after passing a very indifferent night in a narrow and deeply-laden canoe. Having been for several weeks accustomed to mountain scenery, gloomy forests, and rainy weather, they were struck by the bareness of the soil, the clearness of the sky, and the mass of reflected light by which the neighbourhood of Cumana is characterized. At sunrise, they saw the *zamuro* vultures (*Vultur aura*) perched on the cocoa-trees in large flocks. These birds go to roost long before night, and do not quit their place of repose until after the heat of the solar rays is felt. The same idleness, as it were, is indulged by the trees with pinnate leaves, such as the mimosas and tamarinds, which close these organs half an hour before the sun goes down, and unfold them in the morning only after he has been some time visible. In our climates the leguminous plants open their leaves during the morning twilight. Humboldt seems to think that the humidity deposited upon the parenchyma by the refrigeration of the foliage, which is the effect of the nocturnal radiation, prevents the action of the first rays of the sun upon them.

CHAPTER IX.

Indians of New Andalusia.

Physical Constitution and Manners of the Chaymas—Their Languages—American Races.

It is the custom of Humboldt, in his " Journey to the Equinoctial Regions," to stand still after an excursion, reflect, and present to his readers the result of his inquiries on any subject that has fixed his attention. For example, on concluding the narrative of his visit to the Chayma missions, he gives a general account of the aborigines of New Andalusia, of which an abridgment is here offered.

The north-eastern part of Equinoctial America, Terra Firma, and the shores of the Orinoco, resemble, in the multiplicity of the tribes by which they are inhabited, the defiles of Caucasus, the mountains of Hindookho, and the northern extremity of Asia, beyond the Tungooses and the Tartars of the mouth of the Lena. The barbarism which prevails in these various regions is perhaps less owing to an original absence of civilisation than to the effects of a long debasement; and if every thing connected with the first population of a continent were known, we should probably find that savages are merely tribes banished from society and driven into the forests. At the commencement of the conquest of America, the natives were collected into large bodies only on the ridge of the Cordilleras and the coast opposite to Asia, while the vast savannahs, and the great plains covered by forests and intersected by rivers, presented wandering tribes, separated by differences of language and manners.

In New Andalusia, Cumana, and New Barcelona, the aborigines still form fully one-half of the scanty population. Their number may be about 60,000, of which 24,000 inhabit the first of these provinces. This amount appears large when we refer to the hunting tribes of North America, but seems the reverse when we look to those districts of New Spain where agriculture has been followed for more than eight centuries. Thus, the intendancy of Oaxaca, which forms part of the old Mexican empire, and which is one-third smaller than the two provinces of Cumana and Barcelona, contains more than 400,000 of the original race. The Indians of Cumana do not all live assembled in the missions, some being found dispersed in the neighbourhood of towns along the coasts. The stations of the Arragonese Capuchins contain 15,000, almost all of the Chayma tribe. The villages, however, are less crowded than in the province of Barcelona, their indigenous population being only between five and six hundred; whereas, more to the west, in the establishments of the Franciscans of Piritoo, there are towns of 2000 or 3000 inhabitants. Besides the 60,000 natives of the provinces of Cumana and Barcelona, there are some thousands of Guaraounoes who have preserved their independence in the islands at the mouth of the Orinoco. Excepting a few families, there are no wild Indians in New Andalusia.

The term *wild* or *savage*, Humboldt says he uses

with regret, because it implies a difference of cultivation which does not always exist between the *reduced* or *civilized* Indian, living in the missions, and the free or independent Indian. In the forests of South America there are tribes which dwell in villages, rear plantains, cassava, and cotton, and are scarcely more barbarous than those in the religious establishments, who have been taught to make the sign of the cross. It is an error to consider all the free natives as wandering hunters; for agriculture existed on the continent long before the arrival of the Europeans, and still exists between the Orinoco and the Amazons, in districts to which they have never penetrated. The system of the missions has produced an attachment to landed property, a fixed residence, and a taste for quiet life; but the baptized Indian is often as little a Christian as his heathen brother is an idolater,—both discovering a marked indifference for religious opinions, and a tendency to worship nature.

There is no reason to believe, that in the Spanish colonies the number of Indians has diminished since the conquest. There are still more than six millions of the copper-coloured race in both Americas; and although tribes and languages have been destroyed or blended in those colonies, the natives have in fact continued to increase. In the temperate zone the contact of Europeans with the indigenous population becomes fatal to the latter; but in South America the result is different, and there they do not dread the approach of the whites. In the former case a vast extent of country is required by the Indians, because they live by hunting; but in the

G

latter a small piece of ground suffices to afford subsistence for a family.

In these provinces the Europeans advance slowly; and the religious orders have founded establishments between the regions inhabited by them and those possessed by the independent Indians. The missions have no doubt encroached on the liberty of the natives, but they have generally been favourable to the increase of the population. As the preachers advance into the interior the planters invade their territory; the whites and the castes of mixed breed settle among the Indians; the missions become Spanish villages; and finally, the old inhabitants lose their original manners and language. In this way civilisation advances from the coasts towards the centre of the continent.

New Andalusia and Barcelona contain more than fourteen tribes of Indians. Those of the former are the Chaymas, Guayquerias, Pariagotoes, Quaquas, Aruacas, Caribs, and Guaraounoes; and those of the latter, the Cumanagatoes, Palenkas, Caribs, Piritoos, Tomoozas, Topocuares, Chacopatas, and Guarivas. The precise number of the Guaraounoes, who live in huts elevated on trees at the mouth of the Orinoco, is not known. There are two thousand Guayquerias in the suburbs of Cumana and the peninsula of Araya. Of the other tribes the Chaymas of the mountains of Caripe, the Caribs of New Barcelona, and the Cumanagatoes of the missions of Piritoo, are the most numerous. The language of the Guaraounoes, and that of the Caribs, Cumanagatoes, and Chaymas, are the most general, and seem to belong to the same stock.

Although the Indians attached to the missions

are all agriculturists, cultivate the same plants, build their huts in the same manner, and lead the same kind of life, yet the shades by which the several tribes are distinguished remain unchanged. There are few of these villages in which the families do not belong to different tribes, and speak different languages. The missionaries have, indeed, prohibited the use of various practices and ceremonies, and have destroyed many superstitions; but they have not been able to alter the essential character common to all the American races, from Hudson's Bay to the Straits of Magellan. The instructed Indian, more secure of subsistence than the untamed native, and less exposed to the fury of hostile neighbours or of the elements, leads a more monotonous life, possesses the mildness of character which arises from the love of repose, and assumes a sedate and mysterious air; but the sphere of his ideas has received little enlargement, and the expression of melancholy which his countenance exhibits is merely the result of indolence.

The Chaymas, of whom more than fifteen thousand inhabit the Spanish villages, and who border on the Cumanagatoes toward the west, the Guaraounoes toward the east, and the Caribs toward the south, occupy part of the elevated mountains of the Cocollar and Guacharo, as also the banks of the Guarapiche, Rio Colorado, Areo, and the Cano of Caripe. The first attempt to reduce them to subjection was made in the middle of the seventeenth century by Father Francisco of Pamplona, a person of great zeal and intrepidity. The missions subsequently formed among these people suffered greatly in 1681, 1697, and 1720, from the invasions of the Caribs; while during six years subsequently to 1730, the population was diminished by the ravages of the small-pox.

The Chaymas are generally of low stature, their ordinary height being about five feet two inches; but their figures are broad and muscular. The colour of the skin is a dull brown inclining to red. The expression of the countenance is sedate and somewhat gloomy; the forehead is small and retiring; the eyes sunk, very long and black, but not so small or oblique as in the Mongolian race; the eyebrows slender, nearly straight, and black or dark-brown, and the eyelids furnished with very long lashes; the cheekbones are usually high; the hair straight; the beard almost entirely wanting, as in the same people, from whom, however, they differ essentially in having the nose pretty long. The mouth is wide, the lips broad but not prominent, the chin extremely short and round, and the jaws remarkable for their strength. The teeth are white and sound, the toothach being a disease with which they are seldom afflicted. The hands are small and slender, while the feet are large and the toes possessed of an extraordinary mobility. They have so strong a family look, that on entering a hut it is often difficult, among grown up persons, to distinguish the father from the son. This is attributable to the circumstance of their only marrying in their own tribe, as well as to their inferior degree of intellectual improvement; the differences between uncivilized and cultivated man being similar to those between wild and domesticated animals of the same species.

As they live in a very warm country they are excessively averse to clothing. In spite of the remonstrances of the monks, men and women remain naked while within their houses; and, when they go out, wear only a kind of cotton gown scarcely reaching to the knees. The dress of the men has sleeves, while that of the women and boys has none, the arms, shoulders, and upper part of the breast being uncovered. Till the age of nine the girls are allowed to go to church naked. The missionaries complain that the feeling of modesty is very little known to the younger of the sex. The women are not handsome; but the maidens have a kind of pleasant melancholy in their looks. No instances of natural deformity occurred to the travellers. Humboldt remarks, that deviations from nature are exceedingly rare among certain races of men, especially such as have the skin highly coloured; an effect which he does not ascribe solely to a luxurious life or the corruption of morals, but rather imagines that the immunity enjoyed by the American Indians arises from hereditary organization. The custom of marrying at a very early age, which depends upon the same circumstance, is stated to be no way detrimental to population. It occurs in the most northern parts of the continent as well as in the warmest, and, therefore, is not dependent upon climate.

They have naturally very little hair on the chin, and the little that appears is carefully plucked out. This thinness of the beard is common to the American race, although there are tribes, such as the Chipeways and the Patagonians, in which it assumes respectable dimensions.

The Chaymas lead a very regular and uniform life. They go to bed at seven and rise at half after four. The inside of their huts is kept very clean, and their hammocks, utensils, and weapons, are arranged in the greatest order. They bathe every day, and, being generally naked, are thus exempted from the filth principally caused by clothing. Besides their cabin in the village, they usually have a smaller one, covered with palm or plantain leaves, in some solitary place in the woods, to which they retire as often as they can; and so strong is the desire among them of enjoying the pleasures of savage life, that the children sometimes wander entire days in the forests. In fact the towns are often almost wholly deserted. As in all semi-barbarous nations, the women are subjected to privation and suffering, the hardest labour falling to their share.

The Indians learn Spanish with extreme difficulty; and even when they perfectly understand the meaning of the words, are unable to express the most simple ideas in that language without embarrassment. They seem to have as little capacity for comprehending any thing belonging to numbers; the more intelligent counting in Spanish with the appearance of great effort only as far as thirty, or perhaps fifty, while in their own tongue they cannot proceed beyond five or six. The construction of the American dialects is so different from that of the several classes of speech derived from the Latin, that the Jesuits employed some of the more perfect among the former instead of their own; and had this system been generally followed the greatest benefit would have resulted from it. The Chayma appeared to Humboldt less agreeable to the ear than that of the other South American tribes.

The Pariagotoes, or Parias, formerly occupied the coasts of Berbice and Essequibo, the peninsula of Paria, and the plains of Piritoo and Parima.

OTHER NATIVE TRIBES.

Little information, however, is furnished respecting them.

The Guaraounoes are dispersed in the delta of the Orinoco, and owe their independence to the nature of their country. In order to raise their houses above the inundations of the river, they support them on the trunks of the mangrove and mauritia palm. They make bread of the flour obtained from the pith of the latter tree. Their excellent qualities as seamen, their perfect knowledge of the mouths and inosculations of that magnificent stream, and their great number, give them a certain degree of political importance. They run with great address on marshy ground, where the whites, the negroes, or other Indian tribes, will not venture; and this circumstance has given rise to the idea of their being specifically lighter than the rest of the natives.

The Guayquerias are the most intrepid fishermen of these countries, and are the only persons well acquainted with the great bank that surrounds the islands of Coche, Margarita, Sola, and Testigos. They inhabit Margarita, the peninsula of Araya, and a suburb of Cumana.

The Quaquas, formerly a very warlike tribe, are now mingled with the Chaymas attached to the missions of Cumana, although their original abode was on the banks of the Assiveru.

The Cumanagatoes, to the number of more than twenty thousand, subject to the Christian stations of Piritoo, live westward of Cumana, where they cultivate the ground. At the beginning of the sixteenth century they inhabited the mountains of the Brigantine and Parabolota.

The Caribbees of these countries are part of the remnant of the great Carib nation.

The natives of America may be divided into two great classes. To the first belong the Esquimaux of Greenland, Labrador, and Hudson's Bay, and the inhabitants of Behring's Straits, Alaska, and Prince William's Sound. The eastern and western branches of this great family, the Esquimaux proper and the Tschougages, are united by the most intimate similarity of language, although separated to the immense distance of eight hundred leagues. The inhabitants of the north-east of Asia are evidently of the same stock. Like the Malays, this hyperborean nation resides only on the seacoast. They are of smaller stature than the other Americans, lively and loquacious. Their hair is straight and black; but their skin is originally white, in which respect they essentially differ from the other class.

The second race is dispersed over the various regions of the continent, from the northern parts to the southern extremity. They are of larger size, more warlike, and more taciturn, and differ in the colour of their skin. At the earliest age it has more or less of a coppery tinge in most of the tribes, while in others the children are fair, or nearly so; and certain tribes on the Orinoco preserve the same complexion during their whole life. Humboldt is of opinion that these differences in colour are but slightly influenced by climate or other external circumstances, and endeavours to impress the idea that they depend on the original constitution.

CHAPTER X.

Residence at Cumana.

Residence at Cumana—Attack of a Zambo—Eclipse of the Sun—Extraordinary Atmospherical Phenomena—Shocks of an Earthquake—Luminous Meteors.

OUR travellers remained a month longer at Cumana. As they had determined to make a voyage on the Orinoco and Rio Negro, preparations of various kinds were necessary; and the astronomical determination of places being the most important object of this undertaking, it was of essential advantage to observe an eclipse of the sun which was to happen in the end of October.

On the 27th, the day before the obscuration, they went out in the evening, as usual, to take the air. Crossing the beach which separates the suburb of the Guayquerias from the landing-place, they heard the sound of footsteps behind, and on turning saw a tall Zambo, who, coming up, flourished a great palm-tree bludgeon over Humboldt's head. He avoided the stroke by leaping aside; but Bonpland was less fortunate, for, receiving a blow above the temple, he was felled to the ground. The former assisted his companion to rise, and both now pursued the ruffian, who had run off with one of their hats, and on being seized, drew a long knife from his trousers. In the mean time some Biscayan merchants, who were walking on the shore, came to their assistance; when the Zambo, seeing himself surrounded, took to his heels and sought refuge in a cowhouse, from which he was led to prison. The inhabitants showed the warmest concern for the strangers, and although Bonpland had a fever during the night he speedily recovered. The object of the Zambo, who soon afterwards succeeded in escaping from the castle of San Antonio, was never satisfactorily made out.

Notwithstanding this untoward accident Humboldt was enabled to observe the eclipse. The days which preceded and followed it displayed very remarkable atmospheric phenomena. It was what is called winter in those countries. From the 10th of October to the 3d of November a reddish vapour rose in the evening, and in a few minutes covered the sky. The hygrometer gave no indication of humidity. The diurnal heat was from 82·4° to 89·6°. Sometimes in the midst of the night the mist disappeared for a moment, when clouds of a brilliant whiteness formed in the zenith, and extended towards the horizon. On the 18th of October they were so transparent that they did not conceal stars even of the fourth magnitude, and the spots of the moon were very clearly distinguished. They were arranged in masses at equal distances, and seemed to be at a prodigious height. From the 28th of October to the 3d of November the fog was thicker than it had yet been. The heat at night was stifling, although the thermometer indicated only 78·8°. The evening breeze was no longer felt; the sky appeared as if on fire, and the ground was every where cracked and dusty. On the 4th of November, about two in the afternoon, large clouds of extraordinary blackness enveloped the mountains of the Brigantine and Tataraqual, extending gradually to the zenith. About four, thunder was heard overhead, but at an immense height, and with a dull and often interrupted sound. At the moment of the strongest electric explosion, two shocks of an earthquake, separated by an interval of fifteen seconds, were felt. The people in the streets filled the air with their cries. Bonpland, who was examining plants, was nearly thrown on the floor, and Humboldt, who was lying in his hammock, felt the concussion strongly. Its direction was from north to south. A few minutes before the first, there was a violent gust of wind followed by large drops of rain. The sky remained cloudy, and the blast was succeeded by a dead calm which continued all night. The setting of the sun presented a scene of great magnificence. The dark atmospheric shroud was rent asunder close to the horizon, and the sun appeared at 12° of altitude on an indigo ground, its disk enormously enlarged and distorted. The clouds were gilded on the edges, and bundles of rays reflecting the most brilliant prismatic colours extended over the heavens. About nine in the evening there was a third shock, which, although much slighter, was evidently attended with a subterranean noise. The barometer was a little lower than usual, but the progress of the horary variations was in no way interrupted. In the night, between the 3d and 4th of November, the red vapour was so thick that the place of the moon could be distinguished only by a beautiful halo, 20° in diameter.

Scarcely twenty-two months had elapsed since

the almost total destruction of Cumana by an earthquake; and as the people look on the vapours, and the failure of the breeze during the night, as prognostics of disaster, the travellers had frequent visits from persons desirous of knowing whether their instruments indicated new shocks on the morrow. On the 5th, precisely at the same hour, the same phenomena recurred, but without any agitation; and the gust accompanied by thunder returned periodically for five or six days.

This earthquake, being the first that Humboldt ever felt, made a strong impression upon him; but scenes of this kind afterwards became so familiar as to excite little apprehension. It appeared to have a sensible influence on the magnetical phenomena. Soon after his arrival on the coasts of Cumana, he found the dip of the needle 43·53° of the centesimal division. On the 1st of November it was 43·65°. On the 7th, three days after the concussion, he was astonished to find it no more than 42·75°, or 90 centesimal degrees less. A year later, on his return from the Orinoco, he still found it 42·80°, though the intensity of the magnetic forces remained the same after as before the event under consideration, being expressed by 229 oscillations in ten minutes of time. On the 7th November he observed the magnetic variation to be 4° 13' 50" E.

The reddish vapour which appeared about sunset ceased on the 7th of November. The atmosphere then assumed its former purity; and the night of the 11th was cool and extremely beautiful. Towards morning a very extraordinary display of luminous meteors was observed in the east by M. Bonpland, who had risen to enjoy the freshness of the air in the gallery. Thousands of fire-balls and falling-stars succeeded each other during four hours, having a direction from north to south, and filling a space of the sky extending from the true east 30 degrees on either side. They rose above the horizon at E.N.E. and at E., described arcs of various sizes, and fell toward S., some attaining a height of 40°, and all exceeding 25° or 30°. No trace of clouds was to be seen, and a very slight easterly wind blew in the lower regions of the atmosphere. All the meteors left luminous traces from five to ten degrees in length, the phosphorescence of which lasted seven or eight seconds. The fire-balls seemed to explode, but the largest disappeared without scintillation; and many of the falling-stars had a very distinct nucleus, as large as the disk of Jupiter, from which sparks were emitted. The light occasioned by them was white,—an effect which must be attributed to the absence of vapours; stars of the first magnitude having, within the tropics, a much paler hue at their rising than in Europe.

As the inhabitants of Cumana leave their houses before four, to attend the first morning mass, most of them were witnesses of this phenomenon, which gradually ceased soon after, although some were still perceived a quarter of an hour before sunrise.

The day of the 12th November was exceedingly hot, and in the evening the reddish vapour reappeared in the horizon, and rose to the height of 14°. This was the last time it was seen that year.

The researches of M. Chladni having directed the attention of the scientific world to fire-balls and falling-stars at the period of Humboldt's departure from home, he did not fail to inquire, during his journey from Caraccas to the Rio Negro, whether the meteors of the 12th November had been seen. He found that they had been observed by various individuals in places very remote from each other; and on returning to Europe was astonished to find that they had been seen there also. The following is a brief account of the facts relating to these phenomena:—1st, The luminous meteors were seen in the E. and E.N.E. at 40° of elevation, from 2 to 6 A.M., at Cumana, in lat. 10° 27′ 52″, long. 66° 30′; at Porto Cabello, in lat. 10° 6′ 52″, long. 67° 5′; and on the frontiers of Brazil, near the equator, in long. 70° west. 2dly, The Count de Marbois observed them in French Guiana, lat. 4° 56′, long. 54° 35′. 3dly, Mr Ellicot, astronomer to the United States, being in the Gulf of Florida on the 12th November, saw an immense number of meteors, some of which appeared to fall perpendicularly; and the same phenomenon was perceived on the American continent as far as lat. 30° 42′. 4thly, In Labrador, in lat. 56° 55′, and lat. 58° 4′; in Greenland, in latitudes 61° 5′ and 64° 14′, the natives were frightened by the vast quantity of fire-balls that fell during twilight, some of them of great size. 5thly, In Germany, Mr Zeissing, vicar of Itterstadt near Weimar, in lat. 50° 59′, long. 9° 1′ E., observed between 6 and 7 in the morning of the 12th November some falling-stars having a very white light. Soon after reddish streaks appeared in the S. and S.W.; and at dawn the south-western part of the sky was from time to time illuminated by white lightning running in serpentine lines along the horizon.

Calculating from these facts, it is manifest that the height of the meteors was at least 1419 miles;

and as near Weimar they were seen in the S. and S.W., while at Cumana they were observed in the E. and N.E., we must conclude that they fell into the sea between Africa and South America, to the west of the Cape Verd Islands.

Without entering into the learned discussion, which Humboldt submits to his readers, respecting the nature of these luminous bodies, we shall merely observe, that he found falling-stars more frequent in the equinoctial regions than in the temperate zone, and also that they occurred oftener over continents and near certain coasts than on the ocean. He states that on the platform of the Andes, there was observed, upwards of forty years ago, a phenomenon similar to that related above as having occurred at Cumana. From the city of Quito an immense number of meteors was seen rising over the volcano of Cayambo, insomuch that the whole mountain was thought to be on fire. They continued more than an hour, and a religious procession was about to be commenced, when the true nature of the luminous appearance was discovered.

CHAPTER XI.

Voyage from Cumana to Guayra.

Passage from Cumana to La Guayra—Phosphorescence of the Sea—Group of the Caraccas and Chimanas—Port of New Barcelona—La Guayra—Yellow Fever—Coast and Cape Blanco—Road from La Guayra to Caraccas.

HAVING completed the partial investigations which their short residence admitted, and having in some measure become acclimatized, the adventurous philosophers prepared to leave Cumana. Passing by sea to La Guayra, they intended to take up their abode in the town of Caraccas until the rainy season should be over; from thence to traverse the Llanos, or great plains, to the missions of the Orinoco; to go up that river as far as the Rio Negro; and to return to Cumana by Angostura, the capital of Spanish Guiana.

On the 16th November, at eight in the evening, they took their passage in one of the boats which trade between these coasts and the West India islands. They are thirty-two feet long, three feet high at the gunwale, without decks, and generally carry from 200 to 250 quintals (181 to 226 cwts. avoirdupois). Although the sea is very rough from Cape Codera to La Guayra, and these boats have an enormous triangular sail, there had not been an instance for thirty years of the loss of one of them on

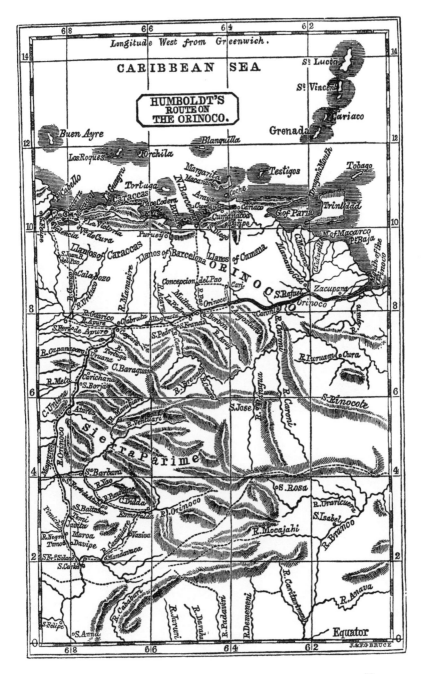

PHOSPHORESCENCE OF THE SEA. 131

the passage from Cumana to Caraccas, so great is the skill of the Guayqueria pilots. They descended the Manzanares with rapidity, delighted with the sight of its marginal cocoa-trees, and the glitter of the thorny bushes covered with noctilucous insects, and left with regret a country in which every thing had appeared new and marvellous. Passing at high water the bar of the river, they entered the Gulf of Cariaco, the surface of which was gently rippled by the evening breeze. In a short time the coasts were recognised only by the scattered lights of the Indian fishermen.

As they advanced toward the shoal that surrounds Cape Arenas, stretching as far as the petroleum springs of Maniquarez, they enjoyed one of those beautiful sights which the phosphorescence of the sea so often displays in tropical climates. When the porpoises, which followed the boat in bands of fifteen or sixteen, struck the surface of the water with their tails, they produced a brilliant light resembling flames. Each troop left behind it a luminous track; and as few sparks were caused by the motion of an oar or of the boat, Humboldt conjectured that the vivid glow produced by these cetaceous animals was owing not to the stroke of their tails alone, but also to the gelatinous matter which envelops their bodies, and which is detached by the waves.

At midnight they found themselves among some rocky islets, rising in the form of bastions, and constituting the group of the Caraccas and Chimanas. Many of these eminences are visible from Cumana, and present the most singular appearances under the effect of mirage. Their height, which is probably not more than 960 feet, seemed much greater when enlightened by the moon, which now shone in a clear sky. The travellers were becalmed in the neighbourhood of these islands, and at sunrise drifted toward Boracha, the largest of them. The temperature had sensibly increased, in consequence of the rocks giving out by radiation a portion of the heat which they had absorbed during the day. As the sun rose, the cliffs projected their lengthened shadows on the ocean, and the flamingoes began to fish in the creeks. These insular spots were all uninhabited; but on one of them, which had formerly been the residence of a family of whites, there were wild goats of a large size and brown colour. The inhabitants had cultivated maize and cassava; but the father, after the death of his children, having purchased two black slaves, was murdered by them. One of the assassins subsequently informed against his accomplice, and at the time of Humboldt's visit was hangman at Cumana.

Proceeding onwards, they anchored for some hours in the road of New Barcelona, at the mouth of the river Neveri, which is full of crocodiles. These animals, especially in calm weather, occasionally make excursions into the open sea,—a fact which is interesting to geologists, on account of the mixture of marine and fresh water organic remains that are occasionally observed in some of the more recent deposites. The port of Barcelona had at that time a very active commerce, arising from the demand in the West Indies for salted provision, oxen, mules, and horses; the merchants of the Havannah being the principal purchasers. Its situation is extremely favourable for this exportation, the animals arriving in three days from the Llanos, while they take more than double that time to reach Cumana, on account of the chain of mountains which they have to cross. Eight thousand mules were embarked at Barcelona, six thousand at Porto Cabello, and three thousand at Carupano, in 1799 and 1800, for the several islands.

Landing on the right bank of the river, they ascended to a small fort, the Morro de Barcelona, built on a calcareous rock, at an elevation of about 400 feet above the sea, but commanded by a much higher hill on the south. Here they observed a very curious geological phenomenon, which recurred in the Cordilleras of Mexico. The limestone, which had a dull, even, or flat conchoidal fracture, and was divided into very thin strata, was traversed by layers of black slaty jasper, with a similar fracture, and breaking into fragments having a parallelopipedal form. It did not exhibit the little veins of quartz so common in Lydian stone, and was decomposed at the surface into a yellowish-gray crust.

Setting sail on the 19th at noon, they found the temperature of the sea at its surface to be $78·6°$; but when passing through the narrow channel which separates the Piritoos, in three fathoms it was only $76·1°$. These islands do not rise more than eight or nine inches above the mean height of the tide, and are covered with long grass. To the westward of the Morro de Barcelona and the mouth of the river Unare, the ocean became more and more agitated as they approached Cape Codera, the influence of which extends to a great distance. Beyond this promontory it always runs very high, although a gale of wind is never felt along this coast. It blew fresh during the night, and on the 20th, at sunrise, they were so far advanced as to be in expectation of doubling the Cape in a few hours; but some of the passengers having suffered from sea-sickness, and the pilot being apprehensive of danger from the privateers stationed near La Guayra, they made for the shore, and anchored at nine o'clock in the Bay of Iliguerota, westward of the Rio Capaya.

On landing, they found two or three huts inhabited by mestizo fishermen, the livid tint of whom, together with the miserable appearance of their children, gave indication of the unhealthy nature of the coast. The sea is so shallow that one cannot go ashore in the smallest boat without wading. The woods come nearly to the beach, which is covered with mangroves, avicennias, manchineel-trees, and *Suriana maritima*, called by the natives *romero de la mar*. Here as elsewhere the insalubrity of the air is attributed to the exhalations from the first of these plants. A faint and sickly smell was perceived, resembling that of the galleries of deserted mines. The temperature rose to 93·2°, and the water along the whole coast acquired a yellowish-brown tint whereever it was in contact with these trees.

Struck by this phenomenon, Humboldt gathered a considerable quantity of branches and roots, with the view of making experiments on the mangrove upon his arrival at Caraccas. The infusion in warm water was of a brown colour, and had an astringent taste. It contained extractive matter and tannin. When kept in contact with atmospheric air under a glass jar for twelve days, the purity of the latter was not perceptibly affected. The wood and roots placed under water were exposed to the rays of

LOW SHORES OF TROPICAL REGIONS. 135

the sun. Bubbles of air were disengaged, which at the end of ten days amounted to a volume of 40 cubic inches. These consisted of azote and carbonic acid, with a trace of oxygen. Lastly, the same substances thoroughly wetted were enclosed with a given volume of atmospheric air in a phial. The whole of the oxygen disappeared. These experiments led him to think that it is the moistened bark and fibre that act upon the atmosphere, and not the brownish water which formed a distinct belt along the coast. Many travellers attribute the smell perceived among mangroves to the disengagement of sulphuretted hydrogen, but no appearance of this kind was observed in the course of these investigations.

"Besides," says Humboldt, "a thick wood covering a muddy ground would diffuse noxious exhalations in the atmosphere, were it composed of trees which in themselves have no deleterious property. Wherever mangroves grow on the margin of the sea, the beach is peopled with multitudes of mollusca and insects. These animals prefer the shade and a faint light; and find shelter from the waves among the closely interlaced roots which rise like lattice-work above the surface of the water. Shells attach themselves to the roots, crustaceous animals nestle in the hollow trunks, the seaweeds which the wind and tide drive upon the shore remain hanging upon the recurved branches. In this manner the maritime forests, by accumulating masses of mud among their roots, extend the domain of the continents; but, in proportion as they gain upon the sea, they scarcely experience any increase in breadth, their very progress becoming the cause of their destruction. The mangroves and the other plants with which they always associate die as the ground dries, and when the salt water ceases to bathe them. Centuries after, their decayed trunks, covered with shells, and half buried in the sand, mark both the route which they have followed in their migrations, and the limit of the land which they have wrested from the ocean."

Cape Codera, seven miles distant from the Bay of Iliguerota, is more imposing on account of its mass than for its elevation, which appeared to be only 1280 feet. It is precipitous on the north, west, and east. Judging from the fragments of rock found along the coast, and from the hills near the town, it is composed of foliated gneiss, containing nodules of reddish felspar, and little quartz. The strata next the bay have the same dip and direction as the great mountain of the Silla, which stretches from Caraccas to Maniquarez in the Isthmus of Araya, and seem to prove that the primitive chain forming that neck of land, after being disruptured or swallowed up by the sea along an extent of 121 miles, reappears at Cape Codera, and runs westward in an unbroken line. Toward the north the cape forms an immense segment of a sphere, and at its foot stretches a tract of low land, known to navigators by the name of the Points of Tutumo and of San Francisco.

The passengers in the boat dreaded the rolling in a rough sea so much, that they resolved to proceed to Caraccas by land, and M. Bonpland, following their example, procured a rich collection of plants. Humboldt, however, continued the voyage, as it seemed hazardous to lose sight of the instruments.

Setting sail at the beginning of the night they

doubled Cape Codera with difficulty, the wind being unfavourable, and the surges short and high. On the 21st of November, at sunrise, they were opposite Curuao, to the west of the Cape. The Indian pilot was frightened at seeing an English frigate only a mile distant; but they escaped without attracting notice. The mountains were every where precipitous, and from 3200 to 4300 feet high, while along the shore was a tract of low humid land, glowing with verdure, and producing a great part of the fruits found so abundantly in the neighbouring markets. The peaks of Niguatar and the Silla of Caraccas form the loftiest summits of this chain. In the fields and valleys the sugar-cane and maize are cultivated. To the west of Caravalleda the declivities along shore are again very steep. After passing this place they discovered the village of Macuto, the black rocks of La Guayra covered with batteries, and in the distance the long promontory of Cabo Blanco, with conical summits of dazzling whiteness.

Humboldt landed at Guayra, and in the evening arrived at Caraccas, four days sooner than his fellow-travellers, who had suffered greatly from the rains and inundations. The former he describes as rather a road than a port, the sea being always agitated, and ships suffering from the action of the wind, the tideways, the bad anchorage, and the worms. The lading is taken in with difficulty. The free mulattoes and negroes, who carry the cocoa on board the ships, are remarkable for their strength. They go through the water up to their middles, although this place abounds in sharks, from which, however, they have in reality nothing to dread. It is singular, that while these animals are dangerous and bloodthirsty at the island opposite the coast of Caraccas, at the Roques, at Buenos Ayres, and at Curassao, they do not disturb persons swimming in the ports of Guayra and Santa Martha. As an analogous fact, Humboldt mentions that the crocodiles of one pool in the Llanos are cowardly, while those of another attack with the greatest fierceness.

The situation of La Guayra resembles that of Santa Cruz in Teneriffe; the houses, which are built on a flat piece of ground about 640 feet broad, being backed by a wall of rock, beyond which is a chain of mountains. The town consists of two parallel streets, and contains 6000 or 8000 inhabitants. The heat is greater than even at Cumana, Porto Cabello, or Coro, the seabreeze being less felt, and the temperature being increased by the radiant caloric emitted by the rocks after sunset.

The examination of the thermometrical observations, made at La Guayra during nine months by Don Joseph Herrera, enabled Humboldt to compare the climate of that port with those of Cumana, Havannah, and Vera Cruz. The result of this comparison was, that the first mentioned is one of the hottest places on the globe; that the quantity of heat which it receives in the course of a year is a little greater than that experienced at Cumana; but that in November, December, and January, the atmosphere cools to a lower point. The mean temperature of the year in these several districts is as follows:—At La Guayra, nearly $82.6°$; at Cumana, $81.2°$; at Vera Cruz, $77.7°$; at Havannah, $78.1°$; while at Rio Janeiro it is $74.5°$; at Santa Cruz in Teneriffe, $71.4°$; at Cairo, $72.3°$; and at Rome, $60.4°$.

At the time of Humboldt's visit to La Guayra,

the yellow fever, or *calentura amarilla*, had been known only two years there, and the mortality had not been very great, as the confluence of strangers was less than at Havannah and Vera Cruz. Some individuals, even creoles and mulattoes, were occasionally taken off by remittent attacks, complicated with bilious symptoms and hemorrhages, and their death often alarmed unseasoned Europeans ; but the disease was not propagated. On the coast of Terra Firma this malignant typhus was known only at Porto Cabello, Carthagena, and Santa Martha. But since 1797 things have changed. The extension of commerce having caused an influx of Europeans and seamen from the United States, the distemper in question soon appeared. It is maintained by some, that it was introduced by a brig from Philadelphia, while others think it took its birth in the country itself, and attribute its origin to a change in the constitution of the atmosphere caused by the overflowings of the Rio de la Guayra, which inundated the town. This fever has since continued its ravages, and has proved fatal not only to troops newly arrived from Spain, but also to those raised far from the coast, in the Llanos between Calabozo and Urituco, a region nearly as hot as La Guayra itself. It scarcely ever passes beyond the ridge of mountains that separates this province from the valley of Caraccas, which has long been exempted from it. The following are the principal pathological facts having reference to this frightful pestilence :—

When a great number of persons, born in a cold climate, arrive at a port in the torrid zone, the insalubrity of which has not been particularly dreaded by navigators, the American typhus (black vomiting, or yellow fever) makes its appearance. These persons, we may add, are not affected by it during the passage; it manifests itself only on the spot. Has the constitution of the atmosphere been changed? asks Humboldt; or, has a new form of disease developed itself in individuals whose excitability is raised to a high pitch? The malady forthwith attacks other Europeans born in warmer countries. Immediate contact does not increase the danger, nor does seclusion diminish it. When the sick are removed to the interior, and especially to cooler and more elevated places, they do not communicate the typhus to the inhabitants. Whenever a considerable diminution of temperature occurs, the distemper usually ceases; but it again begins at the commencement of the hot season, although no ship may have entered the harbour for several months.

The yellow fever disappears periodically at Havannah and at Vera Cruz, when the north winds carry the cold air of Canada towards the Mexican Gulf; but as Porto Cabello, La Guayra, New Barcelona, and Cumana, possess an extreme equality of temperature, it is probable that it will become permanent there. Happily, the mortality has diminished since the treatment has been varied according to the modifications which the disease assumes. In well-managed hospitals, the number of deaths is often reduced to eighteen or fifteen in a hundred; but when the sick are crowded together, the loss increases to one-half or even more.

To the west of La Guayra there are several indentations of the land which furnish excellent anchorage. The coast is granitic, and a great portion of it extremely unhealthy. At Cape Blanco the gneiss passes into mica-slate, containing beds of chlorite-slate, in which garnets and magnetic sand occur. On the road to Catia the chlorite-slate is seen passing into hornblende-slate. At the foot of the promontory the sea throws on the beach rolled fragments of a granular mixture of hornblende and felspar, in which traces of quartz and pyrites are recognised. On the western declivity of that hill the gneiss is covered by a recent sandstone or conglomerate, in which are observed angular fragments of gneiss, quartz, and chlorite, magnetic sand, madrepores, and bivalve shells. The latitude of the Cape is 10° 36′ 45″; that of La Guayra is 10° 36′ 19″, its longitude 67° 5′ 49″.

The road from La Guayra to Caraccas resembles the passages over the Alps; but, as it is kept in tolerable repair, it requires only three hours to go with mules from the port to the capital, and two hours to return. The ascent commences with a ridge of rocks, and is extremely laborious. In the steepest parts the path winds in a zigzag manner. At the *Salto*, or *Leap*, there is a crevice which is passed by a drawbridge, and on the summit of the mountain are fortifications. Half-way is *La Venta* (the *Inn*); beyond which there is a rise of 960 feet to Guayavo, which is not far from the highest part of the route. At the fort of La Cuchilla Humboldt was nearly made prisoner by some Spanish soldiers, whom he, however, contrived to pacify. Round the little inn several travellers were assembled, who were disputing on the efforts that had been made towards obtaining independence; on the hatred of the mulattoes against the free negroes and whites; the wealth of the monks; and on the difficulty of holding slaves in obedience. From Guayavo the road passes over a smooth table-land covered with Alpine plants; and here is seen for the first time the capital, standing nearly 2000 feet lower, in a beautiful valley enclosed by lofty mountains.

The ridges between La Guayra and Caraccas consist of gneiss. On the south side the eminence, which bears the name of Avila, is traversed by veins of quartz, containing rutile in prisms of two or three lines in diameter. The gneiss of the intervening valley contains red and green garnets, which disappear when the rock passes into mica-slate. Near the cross of La Guayra, half a league distant from Caraccas, there were vestiges of blue copper-ore disseminated in veins of quartz, and small layers of graphite. Between the former point and the spring of Sanchorquiz, were beds of bluish-gray primitive limestone, containing mica, and traversed by veins of white calcareous spar. In this deposite were found crystals of pyrites and rhomboidal fragments of sparry iron-ore.

CHAPTER XII.

City of Caraccas and surrounding District.

City of Caraccas—General View of Venezuela—Population—Climate—Character of the Inhabitants of Caraccas—Ascent of the Silla—Geological Nature of the District, and the Mines.

CARACCAS, the capital of the former captain-generalship of Venezuela, is more known to Europeans on account of the earthquakes by which it was desolated than from its importance in a political or commercial point of view. At the present day it is the chief city of a district of the same name, forming part of the republic of Columbia; though, at the time of Humboldt's visit, it was the metropolis of a Spanish colony which contained nearly a million of inhabitants, and consisted of New Andalusia, or the province of Cumana, New Barcelona, Venezuela or Caraccas, Coro, and Maracaybo, along the coast; and in the interior, the provinces of Varinas and Guiana.

In a general point of view Venezuela presents three distinct zones. Along the shore, and near the chain of mountains which skirts it, we find cultivated land; behind this, savannahs or pasturages; and beyond the Orinoco, a mass of forests, penetrable only by means of the rivers by which it is traversed. In these three belts, the three principal stages of civilisation are found more distinct than in almost any other region. We have the life of the wild hunter in the woody district—the pastoral life in the savannahs—and the agricultural in the valleys and plains which descend to various parts of the coast. Missionaries and a few soldiers occupy advanced posts on the southern frontiers. In this section are felt the preponderance of force and the abuse of power. The native tribes are engaged in perpetual hostilities; the monks endeavour to augment the little villages of their missions by availing themselves of the dissensions of the Indians; and the soldiers live in a state of war with the clergy. In the second division, that of the plains and prairies, where food is extremely abundant, little advance has been made in civilisation, and the inhabitants live in huts partly covered with skins. It is in the third district alone, where agriculture and commerce are pursued, that society has made any progress.

In following our travellers through these interesting countries, it is necessary that we lose sight in some measure of the present constitution of the South American states, and view them simply as Spanish provinces. When we seek, says Humboldt, to form a precise idea of those vast regions, which for ages have been governed by viceroys and captains-general, we must fix our attention on several points. We must distinguish the parts of Spanish America that are opposite to Asia, and those that are washed by the Atlantic,—we must observe where the greatest part of the population is placed, whether near the coast or in the interior, or on the table-lands of the Cordilleras,—we must determine the numerical proportions between the natives and other inhabitants, and examine to what race, in each part of the colonies, the greater number of whites belong. The inhabitants of the different districts of the mother-country preserve in some measure their moral peculiarities in the New World, although they have undergone various modifications depending upon the physical constitution of their new abode.

In Venezuela, whatever is connected with an advanced state of civilisation is found along the coast, which has an extent of more than two hundred leagues. It is washed by the Caribbean Sea, a kind of Mediterranean, on the shores of which almost all the European nations have founded colonies, and which communicates at several points with the Atlantic Ocean. Possessing much facility of intercourse with the inhabitants of other parts of America, and with those of Europe, the natives have acquired a great degree of knowledge and opulence.

The Indians constitute a large proportion of the agricultural residents in those places only where the conquerors found regular and long-established governments, as in New Spain and Peru. In the province of Caraccas, for example, the native population is inconsiderable, having been in 1800 not more than one-ninth of the whole, while in Mexico it formed nearly one-half. The black slaves do not exceed one-fifteenth of the general mass, whereas in Cuba they were in 1811 as one to three, and in other West India islands still more numerous. In the Seven United Provinces of Venezuela, there were 60,000 slaves; while Cuba, which has but one-eighth of the extent, had 212,000. The blacks of these countries are so unequally distributed, that in the district of Caraccas alone there were nearly

40,000, of which one-fifth were mulattoes. Humboldt estimates the creoles, or Hispano-Americans, at 210,000 in a population of 900,000, and the Europeans, not including troops, at 12,000 or 15,000.

Caraccas was then the seat of an audiencia, or high court of justice, and one of the eight archbishoprics into which Spanish America was divided. Its population in 1800 was about 40,000. In 1766 great devastation was made by the small-pox, from 6000 to 8000 individuals having perished; but since that period inoculation has become general. In 1812 the inhabitants amounted to 50,000, of which 12,000 were destroyed by the earthquakes; while the political events which succeeded that catastrophe reduced their number to less than twenty thousand.

The town is situated at the entrance of the valley of Chacao, which is ten miles in length, eight and a half miles in breadth, and about 2650 feet above the level of the sea. The ground occupied by it is a steep uneven slope. It was founded by Diego de Losada in 1567. Three small rivers descending from the mountains traverse the line of its direction; it contained eight churches, five convents, and a theatre capable of holding 1500 or 1800 persons. The streets were wide, and crossed each other at right angles; the houses spacious and lofty.

The small extent of the valley, and the proximity of the mountains of Avila and the Silla, give a stern and gloomy character to the scenery, particularly in November and December, when the vapours accumulate towards evening along the high grounds; in June and July, however, the atmosphere is clear and the air pure and delicious. The two rounded summits of the latter are seen from Caraccas, nearly under the same angle of elevation as the Peak of Teneriffe is observed from Orotava. The first half of the ascent is covered with grass; then succeeds a zone of evergreen trees; while above this the rocky masses rise in the form of domes destitute of vegetation. The cultivated region below forms an agreeable contrast to the sombre aspect of the towering ridges which overhang the town, as well as of the hills to the north.

The climate of Caraccas is a perpetual spring, the temperature by day being between 68° and 79°, and by night between 60° and 64°. It is, however, liable to great variations, and the inhabitants complain of having several seasons in twenty-four hours, as well as a too rapid transition from one to another. In January, for example, a night of which the mean heat does not exceed 60° is followed by a day in which the thermometer rises above 71° in the shade. Although in our mild climates oscillations of this kind produce no disagreeable effects, yet in the torrid zone Europeans themselves are so accustomed to uniformity in the temperature, that a difference of a few degrees is productive of unpleasant sensations. This inconvenience is aggravated here by the position of the town in a narrow valley, which is at one time swept by a wind from the coast, loaded with humidity, and depositing its moisture in the higher regions as the warmth decreases; and at another by a dry breeze from the interior, which dissipates the vapours and unveils the mountain-summits. This inconstancy of climate, however, is not peculiar to Caraccas, but is common to the whole equinoctial regions near the tropics. Uninterrupted serenity during a great part of the year prevails only in the low districts adjoining the sea, or on the elevated table-lands of the interior. The intermediate zone is misty and variable.

In this province the sky is generally less blue than at Cumana. The intensity of colour measured by Saussure's cyanometer was commonly 18°, and never above 20°, from November to January, while on the coasts it was from 22° to 25°. The mean temperature is estimated by Humboldt at 68° or 72°. The heat very seldom rises to 84°, and in winter it has been observed to fall as low as 52°. The cold at night is more felt on account of its being usually accompanied by a misty sky. Rains are very frequent in April, May, and June. No hail falls in the low regions of the tropics, but it is seen here every fourth or fifth year.

The coffee-tree is much cultivated in the valley, and the sugar-cane thrives even at a still greater height. The banana, the pine-apple, the vine, the strawberry, the quince, the apple, the peach, together with maize, pulse, and corn, grow in great perfection. But although the atmospheric constitution of this Alpine vale be favourable to diversified culture, it is not equally so to the health of the inhabitants, as the inconstancy of the weather, and the frequent suppression of cutaneous perspiration, give rise to catarrhal affections; and a European, once accustomed to the violent heat, enjoys better health in the low country, where the air is not very humid, than in the elevated and cooler districts.

The travellers remained two months at Caraccas, where they lived in a large house in the upper part of the town, from which they had an extensive view of the mountain-plain, the ridge of the Gallipano, and the summit of the Silla. It was the season of drought, and the conflagrations intended to improve the pasturage produced the most singular effects when seen at night.

They experienced the greatest kindness from all classes of the inhabitants, and more especially from the captain-general of the province, M. de Guevara Vascongelos. Caraccas being situated on the continent, and its population less mutable than that of the islands, the national manners had not undergone so material a change. Notwithstanding the increase of the blacks, says Humboldt, at Caraccas and the Havannah, we seem to be nearer Cadiz and the United States than in any other part of the New World. There was nothing to be seen of the cold and assuming air so common in Europe; on the contrary, conviviality, candour, uniform cheerfulness, and politeness of address, characterized the natives of Spanish origin. The travellers found in several families a taste for instruction, some knowledge of French and Italian literature, and a particular predilection for music. But there was a total deficiency of scientific attainments; nor had the simplest of all the physical sciences, botany, a single cultivator. Previous to 1806 there were no printing-offices in Caraccas.

Believing that in a country which presents such enchanting views, and exhibits such a profusion of natural productions, he should find many persons well acquainted with the surrounding mountains, Humboldt yet failed to discover one individual who had visited the summit of the Silla. But the governor having ordered the proprietor of a plantation to furnish the philosophers with negro guides who knew something of the way, they prepared for the ascent.

As in the whole month of December, the mountain had appeared only five times without clouds, and as at that season two clear days seldom succeed each other, they were advised to choose for their excursion an interval when, the clouds being low, they might hope, by passing through them, to enter into a transparent atmosphere. They spent the night of the 2d January at a coffee-plantation, near a ravine, in which the little river Chacaito formed some fine cascades. At five in the morning they set out, accompanied by slaves carrying their instruments, and about seven reached a promontory of the Silla, connected with the body of the mountain by a narrow dyke. The weather was fine and cool. They proceeded along this ridge of rocks, between two deep valleys covered with vegetation; the large, shining, and coriaceous leaves, illumined by the sun, presenting a very picturesque appearance. Beyond this point the ascent became very steep, the acclivity being often from 32° to 33°. The surface was covered with short grass, which afforded no support when laid hold of, and it was impossible to imprint steps in the gneiss. The persons who had accompanied them from the town were discouraged, and at length retired.

Slender streaks of mist began to issue from the woods, and afforded indications of a dense fog. The familiar loquacity of the negro creoles formed a striking contrast to the gravity of the Indians who had attended the travellers in the missions of Caripe. They amused themselves at the expense of the deserters, among whom was a young Capuchin monk, a professor of mathematics, who had promised to fire off rockets from the top of the mountain, to announce to the inhabitants of Caraccas the success of the expedition.

The eastern peak being the most elevated, they directed their course to it. The depression between the two summits has given rise to the name *Silla*, which signifies a saddle. From this hollow a ravine descends towards the valley of Caraccas. This narrow opening originates near the western dome, and the eastern summit is accessible only by going first to the westward of it, straight over the promontory of the Puerta.

From the foot of the cascade of Chacaito to an elevation of 6395 feet they found only savannahs or pastures, among which were observed two small liliaceous plants with yellow flowers and some brambles. Mixed with the latter they expected to find a wild rose, but were disappointed; nor did they subsequently meet with a single species of that genus in any part of South America.

Sometimes lost in the mist, they made their way with difficulty, and there being no path, they were obliged to use their hands in climbing the steep and slippery ascent. A vein of porcelain-clay, the remains of decomposed felspar, attracted their attention. Whenever the clouds surrounded them the thermometer fell to 53·6°; but when the sky was clear it rose to 69·8°. At the height of 6011 feet they saw in a ravine a wood of palms, which formed a striking contrast with the willows scattered at the bottom of the valley.

After proceeding four hours across the pastures they entered a small forest. The acclivity became less steep, and they observed a profusion of rare and beautiful plants. At the height of 6395 feet the savannahs terminate, and are succeeded by a zone of shrubs, with tortuous branches, rigid leaves, and large purple flowers, consisting of rhododendra, thibaudiæ, andromedæ, vaccinia, and befariæ.

Leaving this little group of Alpine plants they again found themselves in a savannah, and climbed over part of the western dome, to descend into the hollow which separates the two summits. Here the vegetation was so strong and dense, that they were obliged to cut their way through it. On a sudden they were enveloped in a thick mist, and being in danger of coming inadvertently upon the brink of an enormous wall of rocks, which on the north side descends perpendicularly to the depth of more than 6000 feet, were obliged to stop. At this point, however, the negroes who carried their provisions, and who had been detained by the recreant philosopher already mentioned, overtook them, when they made a poor repast, the negroes or the padre having left nothing but a few olives and a little bread. The guides were discouraged, and were with difficulty prevented from returning.

In the midst of the fog the electrometer of Volta, armed with a smoking match, gave very sensible signs of atmospheric electricity, varying frequently from positive to negative, and this, together with the conflict of small currents of air, appeared to indicate a change of weather. It was only two in the afternoon, and they yet entertained some hope of reaching the eastern summit before sunset, and of returning to the hollow separating the two peaks, where they might pass the night. With this view they sent half of their attendants to procure a supply, not of olives but of salt beef. These arrangements were scarcely made when the east wind began to blow violently, and in less than two minutes the clouds disappeared. The obstacles presented by the vegetation gradually diminished as they approached the eastern summit, in order to attain which it was necessary to go close to the great precipice. Hitherto the gneiss had preserved its lamellar structure; but as they climbed the cone of the Silla they found it passing into granite, containing instead of garnets a few scattered crystals of hornblende. In three quarters of an hour they reached the top of the pyramid, which was covered with grass, and for a few minutes enjoyed all the serenity of the sky. The elevation being 8633 feet, the eye commanded a vast range of country. The slope, which extends nearly to the sea, had an angle of 53° 28', though when viewed from the coast it seems perpendicular. Humboldt remarks that a precipice of 6000 or 7000 feet is a phenomenon much rarer than is usually believed, and that a rock of 1600 feet of perpendicular height has in vain been sought for among the Swiss Alps. That of the Silla is partly covered with vegetation, tufts of befariæ and andromedæ appearing as if suspended from the rock.

Seven months had elapsed since they were on the summit of the Peak of Teneriffe, where the apparent horizon of the sea is six leagues farther distant than on the Silla; yet while the boundary-line was seen distinct in the former place, it was completely blended with the air in the latter. The western dome concealed the town of Caraccas; but they distinguished the villages of Chacao and Petare, the coffee-plantations, and the course of the Rio Guayra. While they were examining the part of the sea where the horizon was well defined, and the great chain of mountains in the distant south, a dense fog arose from the plains, and they were obliged to use all expedition in completing their observations.

When seated on the rock, employed in determining the dip of the needle, Humboldt found his hands covered by a species of hairy bee, a little smaller than the honey-bee of Europe. These insects make their nest in the ground, seldom fly, move very slowly, and are apt to use their sting, the guides asserting that they do so only when seized by the legs.

The temperature varied from 52° to 57°, according as the weather was calm or otherwise. The dip of the needle was one centesimal degree less than at Caraccas. The breeze was from the east, which might indicate that the trade-winds extend in this latitude much higher than 9600 feet. The blue of the atmosphere was deeper than on the coasts, Saussure's cyanometer indicating 26·5°, while at Caraccas it generally gave only 18° in fine dry weather. The phenomenon that most struck the travellers was the apparent aridity of the air, which seemed to increase as the mist thickened, the hygrometer retrograding, and their clothes remaining dry.

As it would have been imprudent to remain long in a dense fog, on the brink of a precipice, the travellers descended the eastern dome, and, on regaining the hollow between the two summits, were surprised to find round pebbles of quartz, a phenomenon which perhaps indicates that the mountain has been raised by a power applied from below. Relinquishing their design of passing the night in that, valley, and having again found the path which they had cut through the wood, they soon arrived at the district of resinous shrubs, where they lingered so long collecting plants that darkness surprised them as they entered the savannah. The moon was up, but every now and then obscured by clouds. The guides who carried the instruments slunk off successively to sleep among the cliffs; and it was not until ten that the travellers arrived at the bottom of the ravine, overcome by thirst and fatigue.

During the excursion to the Silla, and in all their walks in the valley of Caraccas, they were very attentive to the indication of ores which they found in the gneiss mountains. In America that rock has not hitherto been found to be very rich in metals, the most celebrated mines of Mexico and Peru being in primitive and transition slate, trap, porphyry, graywacke, and Alpine limestone. In several parts of the region now visited a small quantity of gold was found disseminated in veins of quartz, sulphuretted silver, blue copper-ore, and leadglance; but these deposites did not seem of any importance. In the group of the western mountains of Venezuela, the Spaniards, in 1551, attempted the gold mine of Buria, but the works were soon given up. In the vicinity of Caraccas some had also been wrought, but to no great extent. In short, the mines here afforded little gratification to the cupidity of the conquerors, and were almost totally abandoned, those of Arva, near San Felipe el Fuerte, being the only ones in operation when Humboldt visited the country.

In the course of their investigations the travellers examined the ravine of Tipe, situated in that part of the valley which opens toward Cape Blanco. The first portion of the road was over a barren and rocky soil, on which grew a few plants of *Argemone Mexicana*. On either side of the defile was a range of bare mountains, and at this spot the plain on which the town is built communicates with the coast near Catia by the valleys of Tacagua and Tipe. In the former they found some plantations of maize and plantains, and a very extensive one of cactuses fifteen feet high. They met with several veins of quartz, containing pyrites, carbonated iron-ore, sulphuretted silver, and gray copper. The works that had been undertaken were superficial, and now filled up.

CHAPTER XIII.

Earthquakes of Caraccas.

Extensive Connexion of Earthquakes—Eruption of the Volcano of St Vincent's—Earthquake of the 26th March 1812—Destruction of the City—Ten Thousand of the Inhabitants killed—Consternation of the Survivors—Extent of the Commotions.

THE valley of Caraccas, a few years after Humboldt's visit, became the theatre of one of those physical revolutions which from time to time produce violent alterations upon the surface of our planet; involving the overthrow of cities, the destruction of human life, and a temporary agitation of those elements of nature on which the system of the universe is founded. In the narrative of his Journey to the Equinoctial Regions of the New Continent, he has recorded all that he could collect with certainty respecting the earthquake of the 26th March 1812, which destroyed the city of Caraccas, together with 20,000 inhabitants of the province of Venezuela.

When our travellers visited those countries, they found it to be a general opinion, that the eastern parts of the coasts were most exposed to the destructive effects of such concussions, and that the elevated districts, remote from the shores, were in a great measure secure; but in 1811 all these ideas were proved groundless.

At Humboldt's arrival in Terra Firma, he was

struck with the connexion which appeared between the destruction of Cumana in 1797 and the eruption of volcanoes in the smaller West India islands. A similar principle was manifested in 1812, in the case of Caraccas. From the beginning of 1811 till 1813, a vast extent of the earth's surface, limited by the meridian of the Azores, the valley of the Ohio, the cordilleras of New Grenada, the coasts of Venezuela, and the volcanoes of the West Indies, was shaken by subterranean commotions, indicative of a common agency exerted at a great depth in the interior of the globe. At the period when these earthquakes commenced in the valley of the Mississippi, the city of Caraccas felt the first shock in December 1811, and on the 26th March of 1812 it was totally destroyed.

" The inhabitants of Terra Firma were ignorant of the agitation, which on the one hand the volcano of the island of St Vincent had experienced, and on the other the basin of the Mississippi, where, on the 7th and 8th of February 1812, the ground was day and night in a state of continual oscillation. At this period the province of Venezuela laboured under great drought; not a drop of rain had fallen at Caraccas, or to the distance of 311 miles around, during the five months which preceded the destruction of the capital. The 26th March was excessively hot; the air was calm and the sky cloudless. It was Holy Thursday, and a great part of the population was in the churches. The calamities of the day were preceded by no indications of danger. At seven minutes after four in the evening the first commotion was felt. It was so strong as to make the bells of the churches ring. It lasted from five to six seconds, and was immediately followed by another shock of from ten to twelve seconds, during which the ground was in a continual state of undulation, and heaved like a fluid under ebullition. The danger was thought to be over, when a prodigious subterranean noise was heard, resembling the rolling of thunder, but louder and more prolonged than that heard within the tropics during thunder-storms. This noise preceded a perpendicular motion of about three or four seconds, followed by an undulatory motion of somewhat longer duration. The shocks were in opposite directions, from north to south and from east to west. It was impossible that any thing could resist the motion from beneath upwards, and the undulations crossing each other. The city of Caraccas was completely overthrown. Thousands of the inhabitants (from nine to ten thousand) were buried under the ruins of the churches and houses. The procession had not yet set out; but the crowd in the churches was so great, that nearly three or four thousand individuals were crushed to death by the falling in of the vaulted roofs. The explosion was stronger on the north side of the town, in the part nearest the mountain of Avila and the Silla. The churches of the Trinity and Alta Gracia, which were more than a hundred and fifty feet in height, and of which the nave was supported by pillars from twelve to fifteen feet in diameter, left a mass of ruins nowhere higher than five or six feet. The sinking of the ruins has been so great, that at present hardly any vestige remains of the pillars and columns. The barracks called *El Quartel de San Carlos*, situated further to the north of the church of the Trinity, on the road to the customhouse de la Pastora, almost entirely disappeared. A regiment of troops of the line, which was assembled in it under arms to join in the procession, was, with the exception of a few individuals, buried under this large building. Nine-tenths of the fine town of Caraccas were entirely reduced to ruins. The houses which did not fall, as those of the street of San Juan, near the Capuchin Hospital, were so cracked that no one could venture to live in them. The effects of the earthquake were not quite so disastrous in the southern and western parts of the town, between the great square and the ravine of Caraguata ;—there the cathedral, supported by enormous buttresses, remains standing.

" In estimating the number of persons killed in the city of Caraccas at nine or ten thousand, we do not include those unhappy individuals who were severely wounded, and perished several months after from want of food and proper attention. The night of Holy Thursday presented the most distressing scenes of desolation and sorrow. The thick cloud of dust, which rose above the ruins and darkened the air like a mist, had fallen again to the ground; the shocks had ceased; never was there a finer or quieter night,—the moon, nearly at the full, illuminated the rounded summits of the Silla, and the serenity of the heavens contrasted strongly with the state of the earth, which was strewn with ruins and dead bodies. Mothers were seen carrying in their arms children whom they hoped to recall to life; desolate females ran through the city in quest of a brother, a husband, or a friend, of whose fate they were ignorant, and whom they supposed to have been separated from them in the crowd. The people pressed along the streets, which now could only be distinguished by heaps of ruins arranged in lines.

" All the calamities experienced in the great earthquakes of Lisbon, Messina, Lima, and Riobamba, were repeated on the fatal day of the 26th March 1812. The wounded, buried under the ruins, implored the assistance of the passers by with loud cries, and more than two thousand of them were dug out. Never was pity displayed in a more affecting manner; never, we may say, was it seen more ingeniously active, than in the efforts made to succour the unhappy persons whose groans reached the ear. There was an entire want of instruments adapted for digging up the ground and clearing away the ruins, and the people were obliged to use their hands for the purpose of disinterring the living. Those who were wounded, as well as the patients who had escaped from the hospitals, were placed on the bank of the little river of Guayra, where they had no other shelter than the foliage of the trees. Beds, linen for dressing their wounds, surgical instruments, medicines, in short every thing necessary for their treatment, had been buried in the ruins. During the first days nothing could be procured,— not even food. Within the city water became equally scarce. The commotion had broken the pipes of the fountains, and the falling in of the earth had obstructed the springs which supplied them. To obtain water it was necessary to descend as far as the Rio Guayra, which was considerably swelled, and there were no vessels for drawing it.

" There remained to be performed towards the dead a duty imposed alike by piety and the dread of infection. As it was impossible to inter so many thousands of bodies half buried in the ruins, commissioners were appointed to burn them. Funeral-piles were erected among the heaps of rubbish. This ceremony lasted several days. Amid so many public calamities, the people ardently engaged in the religious exercises which they thought best adapted to appease the anger of heaven. Some walked in bodies chanting funeral-hymns, while others, in a state of distraction, confessed themselves aloud in the streets. In this city was now repeated what had taken place in the province of Quito after the dreadful earthquake of the 4th February 1797. Marriages were contracted between persons who for many years had neglected to sanction their union by the sacerdotal blessing. Children found parents in persons who had till then disavowed them; restitution was promised by individuals who had never been accused of theft; and families, who had long been at enmity, drew together from the feeling of a common evil. But while in some this feeling seemed to soften the heart and open it to compassion, it had a contrary effect on others, rendering them more obdurate and inhumane. In great calamities vulgar minds retain still less goodness than strength; for misfortune acts like the pursuit of literature and the investigation of nature, which exercise their happy influence only upon a few, giving more warmth to the feelings, more elevation to the mind, and more benevolence to the character.

" Shocks so violent as these, which in the space of one minute overthrew the city of Caraccas, could not be confined to a small portion of the continent. Their fatal effects extended to the provinces of Venezuela, Varinas, and Maracaybo, along the coast,

EXTENT OF DAMAGE. 163

and were more especially felt in the mountains of the interior. La Guayra, Mayguetia, Antimana, Baruta, La Vega, San Felipe, and Merida, were almost entirely destroyed. The number of dead exceeded four or five thousand at La Guayra, and at the villa de San Felipe, near the copper-mines of Aroa. The earthquake would appear to have been most violent along a line running from E. N. E. to W. S. W., from Guayra and Caraccas towards the high mountains of Niquitas and Merida. It was felt in the kingdom of New Grenada, from the ramifications of the lofty Sierra of Santa Martha to Santa Fe de Bogota, and Honda on the banks of the Magdalena, 620 miles distant from Caraccas. In all parts it was more violent in the cordilleras of gneiss and mica-slate, or immediately at their base, than in the plains. This difference was particularly remarkable in the savannahs of Varinas and Casanare. In the valleys of Aragua, situated between Caraccas and the town of San Felipe, the shocks were very weak. La Victoria, Maracay, and Valencia, scarcely suffered, notwithstanding the proximity of the capital. At Valecillo, not many leagues distant from Valencia, the ground opened and emitted so great a mass of water that a new torrent was formed. The same phenomenon took place near Porto Cabello. On the other hand, the Lake of Maracaybo underwent considerable diminution. At Coro no commotion was felt, although the town was situated on the coast between other towns which suffered. The fishermen who had passed the day of the 26th March in the island of Orchila, 130 miles N. E. of La Guayra, were not sensible of any shock."

Toward the east of Caraccas the commotions were

very violent, especially beyond Caurimare, in the valley of Capaya, and as far as the meridian of Cape Codera, while they were very feeble on the coasts of New Barcelona, Cumana, and Paria, though these shores are known to have been formerly shaken by volcanic vapours.

Fifteen or eighteen hours after the great catastrophe the ground ceased to be agitated; but subsequently to the 27th the tremblings recommenced, and were accompanied with very loud subterranean noises. Frequently not less than fifteen oscillations were felt in one day. On the 5th April there was an earthquake almost as severe as that of the 12th March. The surface was in continuous undulation during several hours, large masses of earth fell in the mountains, and enormous rocks were detached from the Silla.

While violent agitations were experienced in the valley of the Mississippi, in the island of St Vincent, and in the province of Venezuela, a subterranean noise, resembling an explosion of artillery, was heard at Caraccas, at Calabozo, and on the banks of the Rio Apure, over the space of four thousand square leagues. This sound began at two in the morning of the 30th April, and was as loud on the coast as at the distance of eighty leagues. It was every where taken for the firing of guns. On the same day a great eruption of the volcano of the island of St Vincent took place. This mountain had not ejected lava since 1718, and hardly any smoke was issuing from it, when in May 1811, frequent shocks occurred, and a discharge of ashes, attended with a tremendous bellowing, followed on the 27th April next year. On the 30th the lava flowed, and after a course of four hours reached the sea. The explosions resembled alternate volleys of very large cannon and musketry. As the space between the volcano of St Vincent and the Rio Apure is 725 miles, these were heard at a distance equal to that between Vesuvius and Paris, and must have been propagated by the earth, and not by the air.

After adducing numerous instances of the coincidence of volcanic eruptions and earthquakes, Humboldt endeavours to prove that subterranean communications extend to vast distances, that the phenomena of volcanoes and earthquakes are intimately connected, and that the latter have certain lines of direction.

CHAPTER XIV.

Journey from Caraccas to the Lake of Valencia.

Departure from Caraccas—La Buenavista—Valleys of San Pedro and the Tuy—Manterola—Zamang-tree—Valleys of Aragua—Lake of Valencia—Diminution of its Waters—Hot Springs—Jaguar—New Valencia—Thermal Waters of La Trinchera—Porto Cabello—Cow-tree—Cocoa-plantations—General View of the Littoral District of Venezuela.

LEAVING the city of Caraccas, on their way to the Orinoco, our travellers slept the first night at the base of the woody mountains which close the valley toward the south-west. They followed the right bank of the Rio Guayra, as far as the village of Antimano, by an excellent road, partly scooped out of the rock. The mountains were all of gneiss or mica-slate. A little before reaching that hamlet they observed two large veins of gneiss in the slate, containing balls of granular diabase or greenstone, composed of felspar and hornblende, with garnet disseminated. In the vicinity all the orchards were full of peach-trees covered with flowers. Between Antimano and Ajuntas, they crossed the Rio Guayra seventeen times, and proceeded along the bottom of the valley. The river was bordered by a gramineous plant, the *Gynerium saccharoides*, which sometimes reaches the height of 32 feet, while the huts were surrounded by enormous trees of *Laurus persea*, covered by creepers. They passed the night in a sugar-plantation. In a square house were nearly eighty negroes, lying on skins of oxen spread on the floor, while a dozen fires were burning in the yard, at which people were cooking. A great predilection for the culture of the coffee-tree was entertained in the province. The young plants were chiefly procured by exposing the seeds to germination between plantain-leaves. They were then sown, and produced shoots better adapted to bear the heat of the sun than such as spring up in the shade of the plantations. The tree bears flowers only the second year, and its blossoms last only twenty-four hours. The returns of the third year are very abundant; at an average each plant yielding a pound and a half or two pounds of coffee. Humboldt remarks, that although it is not yet a century since the first trees were introduced at Surinam and in the West Indies, the produce of America already amounts to fifteen millions of piasters, or £2,437,500 sterling.

On the 8th February the travellers set out at sunrise, and after passing the junction of the two small rivers, San Pedro and Macarao, which form the Rio Guayra, ascended a steep hill to the table-land of La Buenavista. The country here had a wild appearance, and was thickly wooded. The road, which was so much frequented that long files of mules and oxen met them at every step, was cut out of a talcose gneiss, in a state of decomposition. Descending from that point, they came upon a ravine, in which a fine spring formed several cascades. Here they found an abundant and diversified vegetation, consisting of arborescent ferns, more than 27 feet high, heliconias, plumerias, brownæ, gigantic figs, palms, and other plants. The brownea, which bears four or five hundred purple flowers in a single thyrsus, reaches the height of fifty or sixty feet.

At the base of the wooded mountain of Higuerota they entered the small village of San Pedro, situated in a basin where several valleys meet. Plantains, potatoes, and coffee, were sedulously cultivated. The rock was mica-slate, filled with garnets, and containing beds of serpentine of a fine green, varied with spots of a lighter tint.

Ascending from the low ground, they passed by the farms of Las Lagunetas and Garavatos, near the latter of which there is a mica-slate rock of a singular form,—that of a ridge, or wall, crowned by a tower. The country is mountainous, and almost entirely uninhabited; but beyond this they entered a fertile district, covered with hamlets and small towns. This beautiful region is the valley of the Tuy, where they spent two days at the plantation of Don Jose de Manterola, on the bank of the river, the water of which was as clear as crystal. Here they observed three species of sugar-cane, the old Creole, the Otaheitan, and the Batavian, which are easily distinguished, and of which the most valuable is the Otaheitan, as it not only yields a third more of juice than the Creole cane, but furnishes a much greater quantity of fuel.

As this valley, like most other parts of the Spanish colonies, has its gold mine, Humboldt was desired to visit it. In the ravine leading to it an enormous tree fixed the attention of the travellers. It had grown on a steep declivity above a house, which it was apprehended it might injure in its fall, should the earth happen to give way. It had therefore been burnt near the root, and cut so as to sink between some large fig-trees, which would prevent it from rolling down. It was eight and a half feet in diameter at the lower end, four feet five inches at the other (the top having been burnt off), and one hundred and sixty feet in length. The rocks were mica-slate passing into talc-slate, and contained masses of bluish granular limestone, together with graphite. At the place where the gold mine was said to have been, they found some vestiges of a vein of quartz; but the subsidence of the earth, in consequence of the rain, rendered it impossible to make any observation. The travellers, however, found a recompense for their fatigues in the harvest of plants which they made in the thick forest abounding in cedrælas, browneas, and fig-trees. They were struck by the woody excrescences, which, as far as twenty feet above the ground, augment the thickness of the latter. Some of these trunks were observed to be twenty-three feet in diameter near the roots.

At the plantation of Tuy, the dip of the needle was 41·6°, and the intensity of the magnetic power was indicated by 228 oscillations in ten minutes. The variation of the former was 4° 30′ N. E. The zodiacal light appeared almost every night with extraordinary brilliancy.

On the 11th, at sunrise, they left the plantation of Manterola, and proceeded along the beautiful banks of the river. At a farm by the way they found a negress more than a hundred years old, seated before a small hut, to enjoy the benefit of the sun's rays, the heat of which, according to her grandson, kept her alive. As they drew near to Victoria the ground became smoother, and resembled the bottom of a lake, the waters of which had been drained off. The neighbouring hills were composed of calcareous tufa. Fields of corn were mingled with crops of sugar-canes, coffee, and plantains. The level of the country above the sea is only from 576 to 640 yards; and, except in the district of Quatro Villas in the island of Cuba, wheat is scarcely cultivated in large quantities in any other part of the equinoctial regions. La Victoria and the neighbouring village of San Matheo yielded 4000 quintals, or 3622 cwt. annually. It is sown in December, and is fit for being cut in seventy or seventy-five days. The grain is large and white, and the average produce is three or four times as much as in Europe. The culture of the sugar-cane, however, is still more productive.

Proceeding slowly on their way, the travellers passed through the villages of San Matheo, Turmero, and Maracay, where every thing was indicative of prosperity. "On leaving the village of Turmero," says Humboldt, "we discover, at the distance of a league, an object which appears on the horizon like a round hillock, or a tumulus covered with vegetation. It is not a hill, however, nor a group of very close trees, but a single tree, the celebrated *Zamang of Guayra*, known over the whole province for the enormous extent of its branches, which form a hemispherical top 614 feet in circumference. The zamang is a beautiful species of mimosa, whose tortuous branches divide by forking. Its slim and delicate foliage is agreeably detached on the blue of the sky. We rested a long while beneath this vegetable arch. The trunk of the Guayra zamang, which grows on the road from Turmero to Maracay, is not more than 64 feet high and $9\frac{1}{2}$ feet in diameter; but its real beauty consists in the general form of its top. The branches stretch out like the spokes of a great umbrella, and all incline towards the ground, from which they uniformly remain twelve or fifteen feet distant. The circumference of the branches or foliage is so regular, that I found the different diameters 205 and 198 feet. One side of the tree was entirely stripped of leaves from the effect of drought, while on the other both foliage and flowers remained. The branches were covered with creeping plants. The inhabitants of these valleys, and especially the Indians, have a great veneration for the Guayra zamang, which the first conquerors seem to have found nearly in the same state as that in which we now see it. Since it has been attentively observed, no change has been noticed in its size or form. It must be at least as old as the dragon-tree of Orotava. Near Turmero and the Hacienda de Cura, there are other trees of the same species, with larger trunks; but their hemispherical tops do not spread so widely."

The valleys of Aragua at this time contained more than 52,000 inhabitants, on a space thirteen leagues in length and two in breadth; making 2000 to a square league, which is almost equal to the densest population of France. The houses were all of masonry, and every court contained cocoa-trees, rising above the habitations; besides wheat, sugar, cacao, cotton, and coffee, indigo is cultivated to a great extent.

In this district the travellers experienced the greatest kindness, more especially from the persons with whom they had associated in Caraccas, and who possessed large estates in these highly-improved and beautiful plains. At the Hacienda de Cura they spent seven very agreeable days, in a small habitation surrounded by thickets, on the Lake of Valencia. Their host, Count Tovar, had begun to let out lands to poor persons, with the view of rendering slaves less necessary to the landholders; and his example was happily followed by other proprietors. Here they lived after the manner of the rich; they bathed twice, slept three times, and made three meals in twenty-four hours.

The valleys of Aragua form a narrow basin between granitic and calcareous mountains of unequal height. On the north they are separated from the coast by the Sierra Mariara, and on the south from the steppes by the chain of Guacimo and Yusma. On the east and west they are bounded by hills of smaller elevation, the rivers from which unite their streams, and are collected in an inland lake, which has no communication with the sea. This body of water, named the Lake of Valencia, and by the Indians called Tacarigua, is larger than the Lake of Neufchatel, but in its general form has more resemblance to that of Geneva. The southern banks are desert, and backed by a screen of high mountains, while the northern shores are decked with the rich cultivation of the sugar-cane, coffee-tree, and cotton. " Paths bordered with cestrum, azedarach, and other shrubs always in flower, traverse the plain and join the scattered farms. Every house is surrounded by a tuft of trees. The ceiba, with large yellow flowers, gives a peculiar character to the landscape, as it unites its branches with those of the purple erythrina. The mixture and brilliancy of the vegetable colours form a contrast to the unvaried tint of a cloudless sky. In the dry season, when the burning soil is covered with a wavy vapour, artificial irrigations keep up its verdure and fecundity. Here and there the granitic rocks pierce the cultivated land, and enormous masses rise abruptly in the midst of the plain, their bare and fissured surfaces affording nourishment to some succulent plants, which prepare a soil for future ages. Often on the summit of these detached hills, a fig-tree or a clusia, with juicy leaves, have fixed their roots in the rock, and overlook the landscape. With their dead and withered branches they seem like signals erected on a steep hill. The form of these eminences reveals the secret of their origin; for when the whole of this valley was filled with water, and the waves beat against the base of the peaks of Mariara, the Devil's Wall, and the coast chain, these rocky hills were shoals or islets."

But the Lake of Valencia is remarkable for other circumstances than its beauties. From a careful examination, Humboldt was convinced that in very remote times, the whole valley from the mountains of Cocuyza to those of Torito and Nirgua, and from the Sierra of Mariara to that of Guigue, Guacimo, and La Palma, had been filled with water. The form of the promontories and their abrupt slopes indicate the shores of an Alpine lake. The same little shells (helicites and valvatæ), which occur at the present day in the Lake of Valencia, are found in layers three or four feet thick in the heart of the country, as far as Turmero and La Concesion near Victoria. These facts prove a retreat of the waters; but no evidence exists that any considerable diminution of them has taken place in recent times, although within the thirty years preceding Humboldt's visit the gradual desiccation of this great basin had excited general attention. This, however, is not dependent upon subterranean channels, as some suppose, but upon the effects of evaporation, increased by the changes operated upon the surface of the country. Forests, by sheltering the soil from the direct action of the sun, diminish the waste of moisture; consequently, when they are imprudently destroyed, the springs become less abundant, or are entirely dried up. Till the middle of the last century, the mountains that surround the valleys of Aragua were covered with woods, and the plains with thickets interspersed with large trees. As cultivation increased the sylvan vegetation suffered; and as the evaporation in this district is excessively powerful, the little rivers were dried up in the lower portion of their course during a great part of the year. The land that surrounds the lake being quite flat and even, the decrease of a few inches in the level of the water exposed a vast extent of ground, and as it retired the planters took possession of the new land.

The idea that the lake will soon entirely disappear, Humboldt treats as chimerical, considering it probable that a period will shortly arrive when the supply of water by the rivers and the evaporation will balance each other. The mean depth is from 77 to 96 feet, and there are some parts not less than 224 or 256 feet. The length is thirty-four and a half miles, and the breadth four or five. The temperature at the surface, in February, was from 73·4° to 74·7°, which was a little lower than the mean temperature of the air.

The Lake of Valencia is covered with beautiful islands to the number of fifteen, some of which are cultivated. It is well stocked with fish, although it furnishes only three kinds, which are soft and insipid. A small crocodile, the bava, which generally attains the length of three or four feet, is very common; but it is remarkable that neither the lake nor any of the rivers which flow into it, have any large alligators, though these animals abound a few leagues off, in the streams that unite with the Apure and Orinoco, or pass directly into the Caribbean Sea. The islands are of gneiss, like the surrounding country. Of the plants which they produce, many have been believed to be peculiar to the district, such as the papaws of the lake, and the tomatoes of the island of Cura. The aquatic vegetation along the shores reminded the travellers of the lakes of Europe, although the species of potamogeton, chara, and equisetum, were peculiar to the New Continent.

Some of the rivers that flow into this fine sheet of water owe their origin to hot springs, of which, however, the travellers were able to examine only those of Mariara and Las Trincheras. In going up the Cura toward its source, the mountains of Mariara are seen advancing into the plain, in the form of an amphitheatre composed of steep rocks, crowned by serrated peaks. The central point is named Rincon del Diablo. These masses are composed of a coarse-grained granite, and are partially covered with vegetation. In the hills toward the east of the Rincon is a ravine containing several small basins, the two uppermost of which are only eight inches in diameter, while the three lower are from two to three feet. Their depth varies from three to fifteen inches, and their temperature is from 133° to 138°. The hot water from these funnels forms a rill, which thirty feet lower has a temperature of only 118·4°. These springs are slightly impregnated with sulphuretted hydrogen gas, the fluid having a thin pellicle of sulphur; while a few plants in the vicinity are crusted with the same substance. To the south of this ravine, in the plain extending to the shores of the lake, is another fountain of the same kind, which issues from a crevice. The water, which is not so hot, collects in a basin fifteen or eighteen feet in diameter and three feet deep, in which the slaves of the neighbouring plantations wash at the end of the day. Here the travellers also bathed, and afterwards found in the surrounding woods a great variety of beautiful plants.

While drying themselves in the sun, after coming out of the pool, a little mulatto approached them, bowing gravely, and making a long speech on the virtues of the water. Showing them his hut, he assured them they should find in it all the conveniences of life; but his attentions ceased the moment he heard they had come merely to satisfy their curiosity, and had no intention to try the efficacy of the baths. They are said to be used with success in rheumatic swellings, old ulcers, and the dreadful affections of the skin called bubas.

On the 21st February, the travellers set out from the Hacienda de Cura for Guacara and New Valencia. As the heat was excessive, they preferred travelling by night. Near the hamlet of **Punta Zamuro**, at the foot of the lofty mountains of **Las Viruelas**, the road was bordered by large mimosas sixty feet in height, and with horizontal branches meeting at a distance of more than fifty yards, so as to form a most beautiful canopy of verdure. The night was gloomy, and the Rincon del Diablo with its serrated cliffs appeared from time to time illuminated by the burning of the savannahs. At a place where the wood was thickest their horses were frightened by the yelling of a large jaguar, which seemed to follow them closely, and which they were informed had roamed among these mountains for three years, having escaped the pursuit of the most intrepid hunters.

They spent the 22d in the house of the Marquis de Foro, at the village of Guacara, a large Indian community; and on the 23d, after visiting Mocundo, an extensive sugar-plantation near it, they continued their journey to New Valencia. They passed a little wood of palms, of the genus *Corypha*, the withered foliage of which, together with the camels feeding in the plain, and the undulating motion of the vapours on the arid soil, gave the landscape quite an African character. The sterility of the land increased as they advanced towards the city, which is said to have been founded in 1555 by Alonzo Diaz Moreno, and contains a population of six or seven thousand individuals. The streets are broad; and as the houses are low, they occupied a large extent of ground. Here the termites or white ants were so numerous, that their excavations resembled subterranean canals, which, being filled with water in rainy weather, became extremely dangerous to the buildings.

On the 26th they set out for the farm of Barbula, to examine a new road that was making from the city to Porto Cabello; and on the 27th visited the hot springs of La Trinchera, three leagues from Valencia. These fountains were so copious as to form a rivulet, which, during the greatest droughts, was two feet deep and eighteen wide. The temperature of the water was 194·5°. Eggs immersed in them were boiled in less than four minutes. They issued from granite, and were strongly impregnated with sulphuretted hydrogen. A sediment of carbonate of lime was deposited, and the most luxuriant vegetation surrounded the basin,—mimosas, clusias, and fig-trees, pushing their roots into the water, and extending their branches over it. Forty feet distant from these remarkable sources there rose others which were of the ordinary temperature. Humboldt remarks, that in all climates people show the same predilection for heat. In Iceland the first Christian converts would be baptized only in the tepid streams of Hecla; and in the torrid zone, the natives flock from all parts to the thermal waters. The river which is formed by the fountains of La Trinchera runs toward the north-east, and near the coast expands to a considerable size.

Descending toward Porto Cabello, the travellers passed through a very picturesque district, beautified by a most luxuriant vegetation and numerous cascades. A stratified coarse-grained granite occurred near the road. The heat became suffocating as they approached the coast, and a reddish vapour veiled the horizon. In the evening they reached the town, where they were kindly received by a French physician, M. Juliac, whose house contained an interesting collection of zoological subjects. This gentleman was principal surgeon to the royal hospital, and was celebrated for his profound acquaintance with the yellow fever. He stated, that when he had treated his patients by bleeding, aperients, and acid drinks, in hospitals where the sick were crowded, the mortality was 33 in 100 among the white creoles, and 65 in 100 among recently-disembarked Europeans; but that since a stimulating treatment, and the use of opium, benzoin, and alcoholic draughts had been substituted for the old debilitating method, the mortality had been reduced to 20 in 100 among Europeans, and 10 among natives.

The heat of Porto Cabello is not so intense as that of La Guayra, the breeze being stronger and more regular, and the air having more room to circulate between the coast and the mountains. The cause of the insalubrity of the atmosphere is therefore to be sought for in the exhalations that arise from the shore to the eastward, where at the beginning of the rainy season tertian fevers prevail, which easily degenerate into the continued typhoid. It has been observed that the mestizoes employed in the salt-works have a yellower skin when they have suffered several successive years from these fevers. The fishermen assert, that the unwholesomeness of the air is owing to the overflowings of the rivers and not to inundations of the sea, and it has been found that the extended cultivation along the banks of the Rio Estevan has rendered them less pestilential.

. The salt-works are similar to those of Araya, near Cumana, but the earth at Porto Cabello contains less muriate of soda. As the employment is very unhealthy, the poorest persons alone engage in it. The defence of the coasts of Terra Firma was maintained at six points, the castle of San Antonio at Cumana, the fortifications of La Guayra, Porto Cabello, Fort St Charles, and Carthagena. Next to Carthagena the most important place is Porto Cabello. The harbour is one of the finest in the world, resembling a basin or little inland lake, opening to the westward by a passage so narrow that only one vessel can anchor at a time, and is defended by batteries. The upper part of it is marshy ground filled with stagnant and putrid water. At the time of Humboldt's visit the number of inhabitants was 9000.

Leaving Porto Cabello on the 1st March at sunrise, our travellers were astonished at the number of boats which they saw laden with fruit for the market. They returned to the valleys of Aragua, and again stopped at the farm of Barbula. Having heard of a tree, the juice of which resembles milk, and is used as an article of food, they visited it, and to their surprise found that the statements which had been made to them with respect to it were correct. It is named the *palo de vaca* or *cow-tree*, and has oblong pointed leaves, with a somewhat fleshy fruit containing one or sometimes two nuts. When an incision is made in the trunk, there issues abundantly a thick glutinous milky fluid, perfectly free from acrimony, and having an agreeable smell. It is drunk by the negroes and free people who work in the plantations, and the travellers took a considerable quantity of it without the least injurious effect. When exposed to the air, this juice presents on its surface a yellowish cheesy substance, in membranous layers, which are elastic, and in five or six days become sour, and afterwards putrefy.

The cow-tree appears to be peculiar to the littoral cordillera, and occurs most plentifully between Barbula and the Lake of Maracaybo.

"Among the many curious phenomena," says Humboldt, "which presented themselves to me in the course of my travels, I confess there were few by which my imagination was so powerfully affected as the cow-tree. All that relates to milk and to the cereal plants inspires us with an interest, which is not merely that of the physical knowledge of things, but which connects itself with another order of ideas and feelings. We can hardly imagine how the human species could exist without farinaceous substances, and without the nutritious fluid which the breast of the mother contains, and which is appropriated to the condition of the feeble infant. The amylaceous matter of the cereal plants,—the object of religious veneration among so many ancient and modern nations,—is distributed in the seeds, and deposited in the roots of vegetables; while the milk which we use as food appears exclusively the product of animal organization. Such are the impressions which we receive in early childhood, and such is the source of the astonishment with which we are seized on first seeing the cow-tree. Magnificent forests, majestic rivers, and lofty mountains clad in perennial snows, are not the objects which we here admire. A few drops of a vegetable fluid impress us with an idea of the power and fecundity of nature. On the parched side of a rock grows a tree with dry and leathery foliage, its large woody roots scarcely penetrating into the ground. For several months in the year its leaves are not moistened by a shower; its branches look as if they were dead and withered; but when the trunk is bored, a bland and nourishing milk flows from it. It is at sunrise that the vegetable fountain flows most freely. At that time the blacks and natives are seen coming from all parts, provided with large bowls to receive the milk, which grows yellow and thickens at its surface. Some empty their vessels on the spot, while others carry them to their children. One imagines he sees the family of a shepherd who is distributing the milk of his flock."

The travellers had resolved to visit the eastern extremity of the cordilleras of New Grenada, where they end in the Paramos of Tirnotes and Niquitas; but learning at Barbula that this excursion would retard their arrival at the Orinoco thirty-five days, they judged it prudent to relinquish it, lest they should fail in the real object of their journey, that of ascertaining by astronomical observations the point at which the Rio Negro and the River of Amazons communicate with the former stream. They therefore returned to Guacara, to take leave of the family of the Marquis del Toro, and pass three days more on the shores of the Lake of Valencia. It happened to be the time of carnival, and all was gayety. The games in which the common people indulged were occasionally not of the most pleasant kind. Some led about an ass laden with water; with which they sprinkled the apartments wherever they found an open window; while others, carrying bags full of the hairs of the *Dolichos pruriens*, which excite great irritation of the skin, blew them into the faces of those who were passing by. From Guacara they

PLANTATIONS OF CACAO.

returned to New Valencia, where they found a few French emigrants, the only ones they saw during five years in the Spanish colonies.

The cacao-plantations have always been considered as the principal source of the prosperity of these countries. The tree (*Theobroma cacao*) which produces this substance is not now found wild in the woods to the north of the Orinoco, and begins to be seen only beyond the cataracts of Atures and Maypures; but it abounds near the Ventuaro, and on the Upper Orinoco. In the plantations it vegetates so vigorously, that flowers spring out even from the woody roots wherever they are left uncovered. It suffers from the north-east winds; and the heavy showers that fall during the winter season, from December to March, are very injurious to it. Great humidity is favourable only when it augments gradually, and continues a long time without interruption. In the dry season, when the leaves and young fruit are wetted by a heavy shower, the latter falls to the ground. For these reasons the cacao-harvest is very uncertain, and the causes of failure are increased by the depredations of worms, insects, birds, and quadrupeds. This branch of agriculture has the disadvantage, moreover, of obliging the new planter to wait eight or ten years for the fruit of his labours, and of yielding an article of very difficult preservation; but it requires a much less number of slaves than most others, one being sufficient for a thousand trees, which at an average yield twelve fanegas annually. It appeared probable, that from 1800 to 1806 the yearly produce of the cacao-plantations of the capitania-general of Caraccas was at least 193,000 fanegas, or 299,200 bushels, of which the province of Caraccas furnished three-fourths. The crops are gathered twice a-year, at the end of June and of December.

Humboldt states, as the result of numerous local estimates, that Europe consumes,—

23,000,000 pounds of cacao, at 12 fr. per cwt.= 27,600,000 fr.
32,000,000 pounds of tea, at 4 fr. per lb.......=128,000,000
140,000,000 pounds of coffee, at 114 fr. per cwt.=159,000,000
450,000,000 pounds of sugar, at 54 fr. per cwt.=243,000,000.

Total value, £23,250,000 sterling, or 558,000,000 fr.

The late wars have had a very injurious effect on the cacao-trade of Caraccas; and the cultivation of this article seems to be gradually declining. It is asserted that the new plantations are not so productive as the old, the trees not acquiring the same vigour, and the harvest being later and less abundant. This is supposed to be owing to exhaustion of the land; but Humboldt attributes it rather to the diminution of moisture caused by cropping.*

In concluding his remarks on the province of Venezuela, our author gives a general view of the soil and metallic productions of the districts of Aroa, Barquesimeto, and Carora. From the Sierra Nevada of Merida, and the Paramos of Niquitao, Bocono, and Las Rosas, the eastern Cordillera of New Grenada decreases so rapidly in height, that between the ninth and tenth degrees of latitude it forms only a chain of hills, which separate the rivers

---

* According to Macculloch, the little use made of this excellent beverage in England may be ascribed to the oppressiveness of the duties with which it has been loaded, and not to its being unsuitable to the public taste. "At this moment (May 1831)," he says, "Trinidad and Grenada cacao is worth in bond, in the London market, from 24s. to 65s. a-cwt.; while the duty is no less than 65s., being nearly 100 per cent. upon the finer qualities, and no less than 230 per cent. upon those that are inferior!"—*Macculloch's Dictionary of Commerce, art. Cacao.*

GEOLOGY OF THE DISTRICT. 185

that join the Apure and the Orinoco from those that flow into the Caribbean Sea or the Lake of Valencia. On this ridge are built the towns of Nirgua, San Felipe, Barquesimeto, and Tocuyo. The ground rises toward the south.

In the cordillera just described, the strata usually dip to the N.W.; so that the waters flow in that direction over the ledges, forming those numerous torrents and rivers, the inundations caused by which are so fatal to the health of the inhabitants from Cape Codera to the Lake of Maracaybo.

Of the streams that descend N.E. toward the coast of Porto Cabello and La Puenta de Hicacos, the most remarkable are the Tocuyo, Aroa, and Taracuy; the valleys of which, were it not for morbid miasmata, would perhaps be more populous than those of Aragua, as the soil is prolific and the waters navigable. In a lateral valley, opening into that of the Aroa, are copper-mines; and in the ravines nearer the sea are similar ores and gold-washings. The total produce of both amounts to a quantity varying from 1087 to 1358 cwts. of excellent metal. Indications of silver and gold have been found in various parts.

The Savannahs or Llanos of Monai and Carora, separated from the great plains of Portuguesa and Calabozo by the mountainous tract of Tocuyo and Migua, although bare and arid, are oppressed with miasmata; and Humboldt seems to think that their insalubrity may be owing to the disengagement of sulphuretted hydrogen gas.

CHAPTER XV.

Journey across the Llanos, from Aragua to San Fernando.

Mountains between the Valleys of Aragua and the Llanos—Their Geological Constitution—The Llanos of Caraccas—Route over the Savannah to the Rio Apure—Cattle and Deer—Vegetation—Calabozo—Gymnoti or Electric Eels—Indian Girl—Alligators and Boas—Arrival at San Fernando de Apure.

From the chain of mountains which borders the Lake of Valencia toward the south, there stretches in the same direction a vast extent of level land, constituting the Llanos or Savannahs of Caraccas; and from the cultivated and populous district of Aragua, embellished with mountains and rivers and teeming with vegetation, one descends into a parched desolate plain, bounded by the horizon. On this route we now accompany our travellers, who on the 6th March left the valleys of Aragua, and keeping along the south-west side of the lake, passed over a rich champaign country covered with calabashes, water-melons, and plantains. The rising of the sun was announced by the howling of monkeys, of which they saw numerous bands moving as in procession from one tree to another. These creatures (the *Simia ursina*) execute their evolutions with singular uniformity. When the boughs of two trees do not touch each other, the leader of the party swings himself by the tail upon the nearest twigs, the rest following in regular succession. The distance to which their howlings may be heard was ascertained by Humboldt to be 1705 yards. The Indians assert that one always chants as leader of the choir; and the missionaries say that when a female is on the point of bringing forth, the howlings are suspended till the moment when the young appears.

The travellers passed the night at the village of Guigue near the lake, where they lodged with an old sergeant, a native of Murcia, who amused them with a recital of the history of the world in Latin, which he had learned among the Jesuits. Leaving this place, they began to ascend the chain of mountains which extends towards La Palma, and from the top of an elevated platform took their last view of the valleys of Aragua. The rock was gneiss with auriferous veins of quartz. Arriving at the hamlet of Maria Magdalena, they were stopped by the inhabitants, who wanted to force their muleteers to hear mass. Seven miles farther on they came to the Villa de Cura, situated in an arid valley almost destitute of vegetation. Here they remained for the night, and joined an assembly of nearly all the residents in the town to admire in a magic-lantern a view of the great capitals of Europe. This place, which contains a population of four thousand, is celebrated for the miracles performed by an image of the Virgin found by an Indian in a ravine.

"Continuing to descend the southern declivity of the range, they passed part of the night of the 11th at the village of San Juan, remarkable for its hot-springs and the singular form of two mountains in the neighbourhood, called the Morros, which rise like slender peaks from a wall of rocks. At two in the morning they continued their journey by Ortiz and Parapara to the Mesa de Paja. The ground over which they travelled forms the ancient shore of the Llanos; and, as the chain has now been traversed, it may be interesting to present a brief view of its geological constitution.

In the Sierra de Mariara, near Caraccas, the rock is coarse-grained granite. The valleys of Aragua, the shores of the Lake of Valencia, its islands, and the southern branch of the coast chain, are of gneiss and mica-slate, which are auriferous. At San Juan some of the rocks were gneiss passing into mica-slate. On the south of this place the gneiss is concealed beneath a deposite of serpentine, which, farther south, passes into or alternates with greenstone. This rock is now the principal one, and in the midst of it rise the Morros of San Juan, composed of crystalline limestone of a greenish-gray colour, and containing masses of dark-blue indurated clay. Behind the Morros is another compact limestone containing shells. The valley that descends from San Juan to the Llanos is filled with trap-rocks lying upon green-slate. Lower down the rocks take a basaltic aspect. Farther south the slates disappear, being concealed under a trap-deposite of varied appearance, but assuming an amygdaloidal character, and on the margin of the plain is seen a formation of clinkstone or porphyry-slate.

The travellers now entered the basin of the Llanos. The sun was almost in the zenith, the ground was at the temperature of 118° or 122°, and the suffocating heat was augmented by the whirls of dust which incessantly arose from the surface of the steril soil.

All around, the plains seemed to ascend into the sky. The horizon in some parts was clear and distinct, while in others it seemed undulating or blended with the atmosphere. The trunks of palm-trees, stripped of their foliage, and seen from afar through the haze, resembled the masts of ships discovered on the verge of the ocean.

In order to give some interest to the narrative of a journey across a tract of so monotonous an aspect, Humboldt presents a general view of the plains of America, contrasted with the deserts of Africa, and the fertile steppes of Asia, of which, however, the most striking points alone can be here taken. There is something awful and melancholy, he says, in the uniform aspect of these savannahs, where every thing seems motionless, and where the shadow of a cloud hardly ever falls for months. He even doubts whether the first sight of the Andes or of the Llanos excites most astonishment; for as mountainous countries have a similarity of appearance, whatever may be the elevation of their summits, the view of a very elevated range is perhaps not so striking as that of a boundless plain, spread out like an ocean, and on all sides mixing with the sky.

It has been said that Europe has its *heaths*, Asia its *steppes*, Africa its *deserts*, and America its *savannahs;* and these great divisions of the globe have been characterized by these circumstances. But as the term heath always supposes the existence of plants of that name, and as all the plains of Europe are not heathy, the description is incorrect. Nor are the steppes of Asia always covered with saline plants, some of them being real deserts; neither are the American Llanos always grassy. Instead of designating the vast levels of these different regions by the nature of the plants which they produce, it seems proper to distinguish them into *deserts*, and *steppes* or *savannahs*, by which terms would be meant plains destitute of vegetation, or covered with grasses or small dicotyledonous plants. The savannahs of North America have been designated by the name of *prairies* or meadows; but the phrase is not very applicable to pastures which are often dry. The Llanos and Pampas of South America are real steppes, displaying a beautiful verdure in the rainy season, but during great droughts assuming the aspect of a desert. The grass is then reduced to powder, the ground cracks, and the alligators and serpents bury themselves in the mud, where they remain in a state of lethargy till they are roused by the showers of spring. On the borders of rivulets, however, and around the little pools of stagnant water, thickets of the Mauritia palm preserve a brilliant verdure, even during the driest part of the year.

The principal characteristic of the savannahs of South America is the entire want of hills. In a space extending to 387 square miles, there is not a single eminence a foot high. These plains, however, present two kinds of inequalities: the *bancos*, consisting of broken strata of sandstone or limestone, which stand four or five feet above the surface; and the *mesas*, composed of small flats or convex mounds, rising gradually to the height of a few yards. The uniform aspect of these flats, the extreme rarity of inhabitants, the fatigue of travelling under a burning sky amid clouds of dust, the continual recession of the horizon, and the successive appearance of solitary palms, make the steppes appear far more extensive than they really are. It has even been imagined that the whole eastern side of South America, from the Orinoco and the Apure to the Plata and the Straits of Magellan, is one great level; but this is not the case. In order to understand their limitations it will be necessary to take a general view of the mountain-ranges.

The cordillera of the coast, where the highest summit is the Silla of Caraccas, and which is connected by the Paramo de las Rosas to the Nevado de Merida, and the Andes of New Grenada, has already been described. A less elevated but much larger group of mountains extends from the mouths of the Guaviare and the Meta, the source of the Orinoco, the Marony, and the Essequibo, toward French and Dutch Guiana. This, which is named the cordillera of Parime, may be followed for a length of 863 miles, and is separated from the Andes of New Grenada by a space 276 miles in breadth. A third chain of mountains, which connects the Andes of Peru with the mountains of Brazil, is the cordillera of Chiguitos, dividing the rivers flowing into the Amazon from the tributaries of the Plata.

These three transverse chains or groups, extending from west to east within the limits of the torrid zone, are separated by level tracts forming the plains of Caraccas or of the Lower Orinoco, the flats of the Amazon and Rio Negro, and those of Buenos Ayres or La Plata. The middle basin, known by the colonists under the name of the *bosques* or *selvas* of the Amazon, is covered with trees; the southern, the *pampas* of Buenos Ayres, with grass; and the northern, the *llanos* of Varinas and Caraccas, with plants of various kinds.

The western coasts of South America are bordered by a wall of mountains, pierced at intervals by volcanic fires, and constituting the celebrated cordillera of the Andes, the mean height of which is 11,830 feet. It extends in the direction of a meridian, sending out two lateral branches, one in lat. 10° north, being that of the coast of Caraccas, the other in lat. 16° and 18° south, forming the cordillera of Chiquitos, and widening eastward in Brazil into vast table-lands. Between these lines is a group of granitic mountains, running from 3° to 7° north latitude, in a direction parallel to the equator, but not united to the Andes. These three chains have no active volcanoes, and none of their summits enter the line of perpetual snow. They are separated by plains, which are closed toward the west and open toward the east; and they are so low, that were the Atlantic to rise 320 feet at the mouth of the Orinoco and 1280 feet at the mouth of the Amazon, more than the half of South America would be covered, and the eastern declivity of the Andes would become a shore of the ocean.

We now accompany the travellers on their route from the northern side of the Llanos to the banks of the Apure, in the province of Varinas. After passing two nights on horseback, they arrived at a little farm called El Cayman, where was a house surrounded by some small huts covered with reeds and skins. They found an old negro who had the management of the farm during his master's absence. Although he told them of herds composed of several thousand cows, they asked in vain for milk, and were obliged to content themselves with some muddy and fetid water drawn from a neighbouring pool, of which they contrived to drink by using a linen cloth as a filter. When the mules were unloaded, they were set at liberty to go and search for water, and the strangers following them came upon a copious reservoir surrounded with palm-trees. Covered with dust and scorched by the sandy wind of the desert, they plunged into the pool, but had scarcely begun to enjoy its coolness when the noise of an alligator floundering in the mud induced them to make a precipitate retreat. Night coming on, they wandered about in search of the farm without succeeding in finding it, and at length resolved to seat themselves under a palm-tree, in a dry spot surrounded by short grass, when an Indian, who had been on his round collecting the cattle, coming up on horseback, was persuaded, though not without difficulty, to guide them to the house. At two in the morning they set off, with the view of reaching Calabozo before noon. The aspect of the country continued the same. There was no moonlight, but the great masses of nebulæ illumined part of the terrestrial horizon as they set out. As the sun ascended, the phenomena of mirage presented themselves in all their modifications. The little currents of air that passed along the ground had so variable a temperature, that in a herd of wild cows some appeared with their legs raised from the surface, while others rested upon it. The objects were generally suspended, but no inversion was observed. At sunrise the plains assumed a more animated appearance; the horses, mules, and oxen, which graze on them in a state of freedom, after having reposed during the night beneath the palms, now assembled in crowds. As the travellers approached Calabozo, they saw troops of small deer feeding in the midst of the cattle. These animals, which are called matacani, are a little larger than the roe of Europe, and have a sleek fawn-coloured pile, spotted with white. Some of them were entirely of the latter hue. Their flesh is good; and their number is so great that a trade in their skins might be carried on with advantage, but the inhabitants are too indolent to engage in any active occupation.

These steppes were principally covered with grasses of the genera *killingia, cenchrus,* and *paspalum,* which at that season scarcely attain a height of nine or ten inches near Calabozo and St Jerome del Pirital, although on the banks of the Apure and Portuguesa they rise to the length of four feet. Along with these were mingled some turneræ, malvaceæ, and mimosæ. The pastures are richest on the banks of the rivers, and under the shade of corypha palms. These trees were singularly uniform in size; their height being from twenty-one to twenty-five feet, and their diameter from eight to ten inches. The wood is very hard, and the fan-like leaves are used for roofing the huts scattered over the plains. A few clumps of a species of rhopala occur here and there.

The philosophers suffered greatly from the heat in crossing the Mesa de Calabozo. Whenever the wind blew, the temperature rose to 104° or 106°, and the air was loaded with dust. The guides advised them to fill their hats with the rhopala leaves, to prevent the action of the solar rays on the head, and from this expedient they derived considerable benefit.

At Calabozo they experienced the most cordial hospitality from the administrator of the Real Hacienda, Don Miguel Cousin. The town, which is situated between the Guarico and the Urituco, has a population of 5000. The principal wealth of the inhabitants consists of cattle, of which it was computed that there were 98,000 in the neighbouring pastures. M. Depons estimates the number in the plains, extending from the mouths of the Orinoco to the Lake of Maracaybo, at 1,200,000 oxen, 180,000 horses, and 90,000 mules; and in the Pampas of Buenos Ayres, it is believed that there are 12,000,000 of cows and 3,000,000 of horses, not including cattle which have no acknowledged owner. In the Llanos of Caraccas, the richer proprietors of the great hatos, or cattle-farms, brand 14,000 head every year, and sell 5000 or 6000. The exportation from the whole capitania-general amounts annually to 174,000 skins of oxen and 11,500 of goats, for the West India Islands alone. This stock was first introduced about 1548 by Christoval Rodriguez. They are of the Spanish breed, and their disposition is so gentle that a traveller runs no risk of being attacked or pursued by them. The horses are also descended from ancestors of the same country, and are generally of a brown colour. There were no sheep in the plains.

Humboldt remarks, that when we hear of the prodigious numbers of oxen, horses, and mules, spread over the plains of America, we forget that in civilized Europe the aggregate amount is not less surprising. According to M. Peuchet, France feeds 6,000,000 of the large horned class; and in the Austrian monarchy, the oxen, cows, and calves, are estimated by Mr Lichtenstein at about 13,400,000.

ε^ΥAt Calabozo, in the midst of the Llanos, the travellers found an electrical apparatus nearly as complete as those of Europe, made by a person who had never seen any such instrument, had received no instructions, and was acquainted with the phenomena of electricity only by reading the Treatise of Sigaud de la Fond, and Franklin's Memoirs. Next to this piece of mechanism, the objects that excited the greatest interest were the electrical eels, or gymnoti, which abound in the basins of stagnant water and the confluents of the Orinoco. The dread of the shocks given by these animals is so great among the common people and Indians, that for some time no specimens could be procured, and one which was at length brought to them, afforded very unsatisfactory results.

On the 19th March, at an early hour, they set off for the village of Rastro de Abaxo, whence they were conducted by the natives to a stream which, in the dry season, forms a pool of muddy water surrounded by trees. It being very difficult to catch the gymnoti with nets, on account of their extreme agility, it was resolved to procure some by intoxicating or benumbing them with the roots of certain plants, which when thrown into the water produce that effect. At this juncture the Indians informed them that they would fish with horses, and soon brought from the savannah about thirty of these animals, which they drove into the pool.

"The extraordinary noise caused by the horses' hoofs makes the fishes issue from the mud, and excites them to combat. These yellowish and livid eels, resembling large aquatic snakes, swim at the surface of the water, and crowd under the bellies of the horses and mules. The struggle between animals of so different an organization affords a very interesting sight. The Indians, furnished with harpoons and long slender reeds, closely surround the pool. Some of them climb the trees, whose branches stretch horizontally over the water. By their wild cries and their long reeds, they prevent the horses from coming to the edge of the basin. The eels, stunned by the noise, defend themselves by repeated discharges of their electrical batteries, and for a long time seem likely to obtain the victory. Several horses sink under the violence of the invisible blows which they receive in the organs most essential to life, and, benumbed by the force and frequency of the shocks, disappear beneath the surface. Others, panting, with erect mane, and haggard eyes expressive of anguish, raise themselves and endeavour to escape from the storm which overtakes them, but are driven back by the Indians. A few, however, succeed in eluding the active vigilance of the fishers; they gain the shore, stumble at every step, and stretch themselves out on the sand, exhausted with fatigue, and having their limbs benumbed by the electric shocks of the gymnoti.

"In less than five minutes two horses were killed. The eel, which is five feet long, presses itself against the belly of the horse, and makes a discharge along the whole extent of its electric organ. It attacks at once the heart, the viscera, and the cæliac plexus of the abdominal nerves. It is natural that the effect which a horse experiences should be more powerful than that produced by the same fish on man, when he touches it only by one of the extremities. The horses are probably not killed but only stunned; they are drowned from the impossibility of rising amid the prolonged struggle between the other horses and eels."

The gymnoti at length dispersed, and approached the edge of the pool, when five of them were taken by means of small harpoons fastened to long cords. A few more were caught towards evening, and there was thus obtained a sufficient number of specimens on which to make experiments. The results of Humboldt's observations on these animals may be stated briefly as follows:—

The gymnotus is the largest electrical fish known, some of those measured by him being from 5 feet 4 inches to 5 feet 7 inches in length. One, 4 feet 1 inch long, weighed $15\frac{3}{4}$ Troy pounds, and its transverse diameter was 3 inches $7\frac{1}{2}$ lines. The colour was a fine olive-green; the under part of the head yellow mingled with red. Along the back are two rows of small yellow spots, each of which contains an excretory aperture for the mucus, with which the skin is constantly covered. The swimming-bladder is of large size, and before it is situated another of smaller dimensions; the former separated from the skin by a mass of fat, and resting upon the electric organs, which occupy more than two-thirds of the fish.

It would be rash to expose one's self to the first shocks of a very large individual,—the pain and numbness which follow in such a case being extremely violent. When in a state of great weakness, the animal produces in the person who touches it a twitching, which is propagated from the hand to the elbow; a kind of internal vibration lasting two or three seconds, and followed by painful torpidity, being felt after every stroke. The electric energy depends upon the will of the creature, and it directs it toward the point where it feels most strongly irritated. The organ acts only under the immediate influence of the brain and heart; for, when one of them was cut through the middle, the fore part of the body alone gave shocks. Its action on man is transmitted and intercepted by the same substances that transmit and intercept the electrical current of a conductor charged by a Leyden jar or a Voltaic pile. In the water the shock can be conveyed to a considerable distance. No spark has ever been observed to issue from the body of the eel when excited.

The gymnoti are objects of dread to the natives, and their presence is considered as the principal cause of the want of fish in the pools of the Llanos. All the inhabitants of the waters avoid them; and the Indians asserted that when they take young alligators and these animals in the same net, the latter never display any appearance of wounds, because they disable their enemies before they are attacked by them. It became necessary to change the direction of a road near Urituco, solely because they were so numerous in a river that they killed many mules in the course of fording it.

On the 24th March the travellers left Calabozo, and advanced southward. As they proceeded they found the country more dusty and destitute of herbage. The palm-trees gradually disappeared. From eleven in the morning till sunset the thermometer kept at 95°. Although the air was calm at the height of eight or ten feet, the ground was swept by little currents which raised clouds of dust. About four in the afternoon, they observed in the savannah a young Indian girl, twelve or thirteen years of age, quite naked, lying on her back, exhausted with fatigue and thirst, and with her eyes, nostrils, and mouth, filled with dust. Her breathing was stertorous, and she was unable to answer the questions put to her. Happily one of the mules was laden with water, the application of which to her face aroused her. She was at first frightened, but by degrees took courage, and conversed with the guides. As she could not be prevailed upon to mount the beasts of burden, nor to return to Urituco, she was furnished with some water; upon which she resumed her way, and was soon separated from her preservers by a cloud of dust.

In the night they forded the Rio Urituco, which is filled with crocodiles remarkable for their ferocity, although those of the Rio Tisnao in the neighbourhood are not at all dangerous. They were shown a hut or shed, in which a singular scene had been witnessed by their host of Calabozo, who, having slept in it upon a bench covered with leather, was awakened early in the morning by a violent shaking, accompanied with a horrible noise. Presently an alligator two or three feet long issued from under the bed, and darted at a dog lying on the threshold, but missing him ran toward the river. When the spot where the bench stood was examined, the dried mud was found turned up to a considerable depth, where the alligator had lain in its state of torpidity or summer sleep. The hut being situated on the edge of a pool, and inundated during part of the year, the animal had no doubt entered at that period and concealed itself in the mire. The Indians often find enormous boas, or water-serpents, in the same lethargic state.

On the 25th March they passed over the smoothest part of the steppes of Caraccas, the Mesa de Pavones. As far as the eye could reach, no object fifteen inches high could be discovered excepting cattle, of which they met some large herds accompanied by flocks of the *crotophaga ani*, a bird of a black colour with olive reflections. They were exceedingly tame, and perched upon the quadrupeds in search of insects.

Wherever excavations had been made, they found the rock to be old red sandstone or conglomerate, in which were observed fragments of quartz, kieselschiefer, and lydian stone. The cementing clay is ferruginous, and often of a very bright red. This formation, which covers an extent of several thousand square leagues, rests on the northern margin of the plains upon transition-slate, and to the south upon the granites of the Orinoco.

After wandering a long time on the desert and pathless savannahs of the Mesa de Pavones, they were agreeably surprised to find a solitary farm-house surrounded with gardens and pools of clear water. Farther on they passed the night near the village of San Geronymo del Guyaval, situated on the banks of the Rio Guarico, which joins the Apure. The ecclesiastic, who was a young man, and had no other habitation than his church, received them in the kindest manner. Crossing the Guarico they encamped in the plain, and early in the morning pursued their way over low grounds which are often inundated. On the 27th they arrived at the Villa de San Fernando, and terminated their journey over the Llanos.

CHAPTER XVI.

Voyage down the Rio Apure.

San Fernando—Commencement of the Rainy Season—Progress of Atmospherical Phenomena—Cetaceous Animals—Voyage down the Rio Apure—Vegetation and Wild Animals—Crocodiles, Chiguires, and Jaguars—Don Ignacio and Donna Isabella—Water-fowl—Nocturnal Howlings in the Forest—Caribe-Fish—Adventure with a Jaguar—Manatees—Mouth of the Rio Apure.

THE town of San Fernando, which was founded only in 1789, is advantageously situated on a large navigable river, the Apure, a tributary of the Orinoco, near the mouth of another stream which traverses the whole province of Varinas, all the productions of which pass through it on their way to the coast. It is during the rainy season, when the rivers overflow their banks and inundate a vast extent of country, that commerce is most active. At this period the savannahs are covered with water to the depth of twelve or fourteen feet, and present the appearance of a great lake, in the midst of which the farm-houses and villages are seen rising on islands scarcely elevated above the surface. Horses, mules, and cows, perish in great numbers, and afford abundant food to the zamuros or carrion vultures, as well as to the alligators. The inhabitants, to avoid the force of the currents, and the danger arising from the trees carried down by them, instead of ascending the course of the rivers, find it safer to cross the flats in their boats.

San Fernando is celebrated for the excessive heat which prevails there during the greater part of the year. The travellers found the white sand of the shores, wherever it was exposed to the sun, to have a temperature of 126·5°, at two in the afternoon. The thermometer, raised eighteen inches above the sand, indicated 109°; and at six feet, 101·7°. The temperature of the air in the shade was 97°. These observations were made during a dead calm, and when the wind began to blow, the heat increased three degrees.

On the 28th March, Humboldt and his companion being on the shore at sunrise, heard the thunder rolling all around, although as yet there were only scattered clouds, advancing in opposite directions toward the zenith. Deluc's hygrometer was at 53°, the thermometer stood at 74·7°, and the electrometer gave no particular indication. As the clouds mustered, the blue of the sky changed to deep azure, and then to gray; and when it was completely overcast the thermometer rose several degrees. Although a heavy rain fell, the travellers remained on the shore to observe the electrometer. When it was held at the height of six feet from the ground, the pith-balls generally separated only a few seconds before the lightning was seen. The separation was four lines. The electric charge remained the same for several minutes, and there were repeated oscillations from positive to negative. Toward the end of the storm the west wind blew with great impetuosity, and when the clouds dispersed the thermometer fell to 71·6°.

Humboldt states, that he enters into these details because Europeans usually confine themselves to a description of the impressions made on their minds by the solemn spectacle of a tropical thunder-storm; and because, in a country where the year is divided into two great seasons of drought and rain, it is interesting to trace the transition from the one to the other. In the valleys of Aragua, he had from the 18th of February observed clouds forming in the evening, and in the beginning of March the accumulation of vesicular vapours became visible. Flashes of lightning were seen in the south, and at sunset Volta's electrometer regularly displayed positive indications, the separation of the pith-balls being from three to four lines. After the 26th of the latter month, the electrical equilibrium of the atmosphere seemed broken, although the hygrometer still denoted great dryness.

The following is an account of the atmospheric phenomena in the inland districts to the east of the cordilleras of Merida and New Grenada, in the Llanos of Venezuela, and the Rio Meta, from the fourth to the tenth degree of north latitude, wherever the rains continue from May to October, and consequently include the period of the greatest heat, which is in July and August:—" Nothing can equal the purity of the atmosphere from December to February. The sky is then constantly without clouds, and should one appear, it is a phenomenon that occupies all the attention of the inhabitants. The breeze from the east and north-east blows with violence. As it always carries with it air of the same temperature, the vapours cannot become visible through refrigeration. Towards the end of Febru-

PHENOMENA IN THE INTERIOR. 205

ary and the beginning of March the blue of the sky is less intense; the hygrometer gradually indicates greater humidity; the stars are sometimes veiled by a thin stratum of vapours; their light ceases to be tranquil and planetary; and they are seen to sparkle from time to time at the height of 20° above the horizon. At this period the breeze diminishes in strength, and becomes less regular, being more frequently interrupted by dead calms. Clouds accumulate towards the south-east, appearing like distant mountains with distinct outlines. From time to time they are seen to separate from the horizon, and traverse the celestial vault with a rapidity which has no correspondence with the feebleness of the wind that prevails in the lower strata of the air. At the end of March the southern region of the atmosphere is illuminated by small electric explosions, like phosphorescent gleams confined to a single group of vapours. From this period the breeze shifts at intervals, and for several hours, to the west and south-west, affording a sure indication of the approach of the rainy season, which on the Orinoco commences about the end of April. The sky begins to be overcast, its azure colour disappears, and a gray tint is uniformly diffused over it. At the same time the heat of the atmosphere gradually increases, and instead of scattered clouds the whole vault of the heavens is overspread with condensed vapours. The howling-monkeys begin to utter their plaintive cries long before sunrise. The atmospheric electricity which, during the period of the greatest drought, from December to March, had been almost constantly in the daytime from 1·7 to 2 lines to Volta's electrometer, becomes extremely variable after March.

During whole days it appears null, and again, for some hours, the pith-balls of the electrometer diverge from three to four lines. The atmosphere, which in the torrid as in the temperate zone is generally in a state of positive electricity, passes alternately, in the course of eight or ten minutes, to the negative state. The rainy season is that of thunderstorms; and yet I have found, from numerous experiments made during three years, that at this season the electric tension is less in the lower regions of the atmosphere. Are thunder-storms the effect of this unequal change of the different superimposed strata of the air? What prevents the electricity from descending towards the earth in a stratum of air which has become more humid since the month of March? At this period the electricity, in place of being diffused through the whole atmosphere, would seem to be accumulated on the outer envelope at the surface of the clouds. According to M. Gay Lussac, it is the formation of the cloud itself that carries the fluid toward the surface. The storm rises in the plains two hours after the sun passes through the meridian, and therefore shortly after the period of the maximum of the diurnal heat in the tropics. In the inland districts it is exceedingly rare to hear thunder at night or in the morning, nocturnal thunder-storms being peculiar to certain valleys of rivers which have a particular climate."

It may be interesting to present a very brief statement of Humboldt's explanation of these phenomena: —The season of rains and thunder in the northern equinoctial zone coincides with the passage of the sun through the zenith of the place, the cessation of the breezes or north-east winds, and the frequency of calms, and furious currents of the atmosphere from the south-east and south-west, accompanied with a cloudy sky. While the breeze from the north-east blows, it prevents the atmosphere from being saturated with moisture. The hot and loaded air of the torrid zone rises and flows off again towards the poles, while inferior currents from these last, bringing drier and colder strata, take the place of the ascending columns. In this manner the humidity, being prevented from accumulating, passes off towards the temperate and colder regions, so that the sky is always clear. When the sun, entering the northern signs, rises towards the zenith, the breeze from the north-east softens, and at length ceases; this being the season at which the difference of temperature between the tropics and the contiguous zone is least. The column of air resting on the equinoctial zone becomes replete with vapours, because it is no longer renewed by the current from the pole; clouds form in this atmosphere, saturated and cooled by the effects of radiation and the dilatation of the ascending air, which increases its capacity for heat in proportion as it is rarefied. Electricity accumulates in the higher regions in consequence of the formation of the vesicular vapours, the precipitation of which is constant during the day, but generally ceases at night. The showers are more violent, and accompanied with electrical explosions, shortly after the maximum of the diurnal heat. These phenomena continue until the sun enters the southern signs, when the polar current is re-established, because the difference between the heat of the equinoctial and temperate regions is daily increasing. The air of the tropics being thus renewed, the rains cease, the vapours are dissolved, and the sky resumes its azure tint.

At San Fernando, Humboldt observed in the river long files of cetaceous animals, resembling the common porpoise. The crocodiles seemed to dislike them, and dived whenever they approached. They were three or four feet long, and appear to be peculiar to the great streams of South America, as he saw some of them above the cataracts of the Orinoco, whither they could not have ascended from the sea.

The rainy season had now commenced, and as the way to that river by land lies across an unhealthy and uninteresting flat, they preferred the longer way by the Rio Apure, and embarked in a large canoe or lancha, having a pilot and four Indians for crew. A cabin was constructed in the stern, of sufficient size to hold a table and benches, and covered with corypha-leaves. They put on board a stock of provisions for a month, while the capuchin missionary, with whom they had lodged during their stay, supplied them with wine, oranges, and tamarinds. Fishing-instruments, fire-arms, and some casks of brandy, for bartering with the natives, were added to their store. On the 30th March, at four in the afternoon, they left San Fernando, accompanied by Don Nicolas Sopo, brother-in-law of the governor of the province. The river abounds in fish, manatees, and turtles, and its banks are peopled by numberless birds, of which the pauxi and guacharaca are the most useful to man. Passing the mouth of the Apurito, they coasted the island of the same name, formed by the Apure and Guarico, and which is seventy-six miles in length. On the banks they saw huts of the Yaruroes, who live by hunting and fishing, and are very skilful in killing jaguars, the skins of which they dispose of in the Spanish villages. The night was passed at Diamante, a small sugar-plantation.

On the 31st a contrary wind obliged them to remain on shore till noon, when they embarked, and as they proceeded found the river gradually widening; one of its banks being generally sandy and barren, the other higher and covered with tall trees. Sometimes, however, it was bordered on both sides by forests, and resembled a straight canal 320 yards in breadth. Bushes of sauso (*Hermesia castaneifolia*) formed along the margins a kind of hedge about four feet high, in which the jaguars, tapirs, and pecaris, had made openings for the purpose of drinking; and as these animals manifest little fear at the approach of a boat, the travellers had the pleasure of viewing them as they walked slowly along the shore, until they disappeared in the forest. When the sauso-hedge was at a distance from the current, crocodiles were often seen in parties of eight or ten, stretched out on the sand motionless, and with their jaws opened at right angles. These monstrous reptiles were so numerous, that throughout the whole course of the river there were usually five or six in view, although the waters had scarcely begun to rise, and hundreds were still buried in the mud of the savannahs. A dead individual which they found was 17 feet 9 inches long, and another, a male, was more than 23. This species is not a cayman or alligator, but a real crocodile, with feet dentated on the outer edge like that of the Nile. The Indians informed them, that scarcely a year passes at San Fernando without two or three persons being drowned by them, and related the history of a young girl of Urituco who, by singular presence of mind, made her escape from one. Finding herself seized and carried into the water, she felt for the eyes of the animal, and thrust her fingers into them; when the crocodile let her loose, after biting off the lower part of her left arm. Notwithstanding the quantity of blood which she lost, she was still able to reach the shore by swimming with the right hand. Mungo Park's guide, Isaaco, effected his preservation from a crocodile by employing the same means. The motions of these animals are abrupt and rapid when they attack an object, although they move very slowly when not excited. In running they make a rustling noise, which seems to proceed from their scales, and appear higher on their legs than when at rest, at the same time bending the back. They generally advance in a straight line, but can easily turn when they please. They swim with great facility, even against the most rapid current. On the Apure they seemed to live chiefly on the chiguires (*Cavia capybara*), which feed in herds on the banks, and are of the size of our pigs. These creatures have no weapons for defence, and are alternately the prey of the jaguars on land and of the crocodiles in the water.

Stopping below the mouth of the Cano de la Figuera, in a sinuosity called La Vuelta del Joval, they measured the velocity of the current at its surface, which was only 3·4 feet in a second. Here they were surrounded by chiguires, swimming like dogs, with the head and neck out of the water. A large crocodile, which was sleeping on the shore

Jaguar, or American Tiger.

in the midst of a troop of these animals, awoke at the approach of the canoe, and moved slowly into the stream without frightening the others. Near the Joval every thing assumed a wild and awful aspect. Here they saw an enormous jaguar stretched beneath the shade of a large zamang or mimosa. It had just killed a chiguire, which it held with one of its paws, while the zamuro-vultures were assembled in flocks around it. It was curious to observe the mixture of boldness and timidity which these birds exhibited, for although they advanced within two feet of the tiger, they instantly shrunk back at the least motion which he made. In order to examine more nearly their manners, the travellers went into the little boat; when the tyrant of the forest withdrew behind the sauso-bushes, leaving his victim, which the vultures in the mean time attempted to devour, but were soon put to flight by his rushing into the midst of them.*

---

* In the province of Tucuman, the common mode of killing the jaguar is to trace him to his lair, by the wool left on the bushes, if he has carried off a sheep, or by means of a dog trained for the purpose. On finding the enemy the gaucho puts himself into a position for receiving him on the point of a bayonet or spear, at the first spring which he makes, and thus waits until the dogs drive him out; an exploit which he performs with such coolness and dexterity that there is scarcely an instance of failure. " In a recent instance, related by our capitaz, the business was not so quickly completed. The animal lay stretched at full length on the ground, like a gorged cat. Instead of showing anger and attacking his enemies with fury, he was playful, and disposed rather to parley with the dogs with good humour than to take their attack in sober earnestness. He was now fired upon, and a ball lodged in his shoulder; on which he sprung so quickly on his watching assailant, that he not only buried the bayonet in his body, but tumbled over the capitaz who held it, and they floundered on the ground together, the man being completely in his clutches. ' I thought,' said the brave fellow, ' I was no longer a capitaz, while I held my arm up to protect my throat, which the animal seemed in the act of seizing; but when I expected to feel his fangs in my flesh, the green fire of his eyes which blazed

Continuing to descend the river, they met with a great herd of chiguires that the tiger had dispersed, and from which he had selected his prey. These animals seemed not to be afraid of men, for they saw the travellers land without agitation, but the sight of a dog put them to flight. They ran so slowly that the people succeeded in catching two of them. It is the largest of the *Glires* or gnawing animals. Its flesh has a disagreeable smell of musk, although hams are made of it in the country, which are eaten during Lent; as this quadruped, according to ecclesiastical zoology, is esteemed a fish.

The travellers passed the night as usual in the open air, although in a plantation, the proprietor of which, a jaguar-hunter, half-naked and as brown as a Zambo, prided himself on being of the European race, and called his wife and daughter, who were as slightly clothed as himself, Donna Isabella and Donna Manuela. Humboldt had brought a chiguire; but his host assured him such food was not fit for white gentlemen like them, at the same time offering him venison. As this aspiring personage had neither house nor hut, he invited the strangers to sling their hammocks near his own, between two trees; which they accordingly did. They soon found reason, however, to regret that they had not obtained better shelter; for after midnight a thunder-storm came on, which wetted them to the skin. Donna Isabella's cat had perched on one of the trees, and fell into a cot, the inmate of which imagined he was attacked by some wild beast, and could hardly be quieted.

---

upon me, flashed out in a moment. He fell on me, and expired at the very instant I thought myself lost for ever.' "—*Captain Andrews' Travels in South America*, vol. i. p. 219.

At sunrise, the lodgers took leave of Don Ignacio and his lady, and proceeded on their voyage. The weather was a little cooler, the thermometer having fallen from 86° to 75°, but the temperature of the river continued at 79° or 80°. One might imagine that on smooth ground, where no eminence can be distinguished, the stream would have hollowed out an even bed for itself; but this is by no means the case; the two banks not opposing equal resistance to the water. Below the Joval the mass of the current is a little wider, and forms a perfectly straight channel, margined on either side by lofty trees. It was here about 290 yards broad. They passed a low island densely covered by flamingoes, roseate spoonbills, herons, and water-hens, which presented a most diversified mixture of colours. On the right bank they found a little Indian mission, consisting of sixteen huts constructed of palm-leaves, and inhabited by a tribe of the Guamoes. These Christians were unable to furnish them with the provisions which they wanted, but hospitably offered them dried fish and water. The night was spent on a bare and very extensive beach. The forest being impenetrable, they had great difficulty in obtaining dry wood to light fires for the purpose of keeping off the wild beasts. But the night was calm, with beautiful moonlight. Finding no tree on the banks, they stuck their oars in the sand, and suspended their hammocks upon them. About eleven there arose in the wood so terrific a noise that it was impossible to sleep. The Indians distinguished the cries of sapajous, alouates, jaguars, cougars, pecaris, sloths, carassows, panakas, and other gallinaceous birds. When the tigers approached the edge of the forest, a dog which the travellers had began to howl and seek refuge under their cots. Sometimes, after a long silence, the cry of the ferocious animal came from the tops of the trees, when it was followed by the sharp and long whistling of the monkeys. Humboldt supposes the noise thus made by the inhabitants of the thicket, at certain hours of the night, to be the effect of some contest that has arisen among them.

On the 2d April they set sail before sunrise. The river was ploughed by porpoises, and the shore crowded with aquatic birds; while some of the latter, perched on the floating timber, were endeavouring to surprise the fish that preferred the middle of the stream. The navigation is rather dangerous, on account of the large trees which remain obliquely fixed in the mud, and the canoe touched several times. Near the island of Carizales, they saw enormous trunks covered with plotuses or darters, and below it observed a diminution of the waters of the river, owing to infiltration and evaporation. Near the Vuelta de Basilio, where they landed to gather plants, they saw on a tree two beautiful jet-black monkeys of an unknown species, and also a nest of iguanas, which was pointed out by the Indians. The flesh of this lizard is very white, and, next to that of the armadillo, is the best food to be found in the huts of the natives. Towards evening it rained, and swallows were seen skimming along the water. They also saw a flock of parrots pursued by hawks. The night was passed on the beach.

On the 3d they proceeded down the river in their solitary course. The sailors caught the fish known in the country by the name of caribe; which, although only four or five inches in length, attacks persons who go into the water, and with its sharp triangular teeth often tears considerable portions of flesh from their legs. When pieces of meat are cast into the river, clouds of these little fishes appear in a few minutes. There are three varieties in the Orinoco; one of which seems to be the *Salmo rhombeus* of Linnæus. At noon they stopped in a desert spot called Algodonal, when Humboldt left his companions and went along the beach to observe a group of crocodiles sleeping in the sun. Some little herons of a white colour were walking along their backs, and even on their heads. As he was proceeding, his eyes directed towards the river, he discovered recent footmarks of a beast of prey, and turning toward the forest, found himself within eighty steps of an enormously large jaguar. Although extremely frightened, he yet retained sufficient command of himself to follow the advice which the Indians had so often given, and continued to walk without moving his arms, making a large circuit toward the edge of the water. As the distance increased he accelerated his pace, and at length, judging it safe to look about, did so, and saw the tiger in the same spot. Arriving at the boat out of breath, he related his adventure to the natives, who seemed to think it nothing extraordinary. In the evening they passed the mouth of the Cano del Manati, so named on account of the vast number of manatees caught there. This aquatic herbivorous animal generally attains the length of ten or twelve feet, and abounds in the Orinoco below the cataracts, the Rio Meta, and the Apure. The flesh, although very savoury and resembling pork, is considered unwholesome; but it is in request during Lent, being classed by the monks among fishes. The fat is used for lamps in the churches, as well as for cooking; while the hide is cut into slips to supply the place of cordage. Whips are also made of it in the Spanish colonies for the castigation of negroes and other slaves. The fires lighted by the boatmen on the shore attracted the crocodiles and dolphins. Two persons kept watch during the night. A jaguar with her cub approached the encampment, but was driven away by the attendants; and soon after the dog was bitten in the nose by a large bat or vampire.

On the 4th they intended to pass the night at Vuelta del Palmito; but as the Indians were going to sling the hammocks they found two tigers concealed behind a tree, and it was judged safer to re-embark and sleep on the island of Apurito. Multitudes of gnats made their appearance regularly at sunset, and covered their faces and hands. On the 5th they were much struck by the diminution the waters of the Apure had undergone, which they attributed chiefly to absorption by the sand and evaporation. It was only from 128 to 170 yards broad, and about twenty feet deep. Humboldt estimates the mean fall of this river at 14 inches in a mile. The canoe touched several times on shoals as they approached the point of junction, and it became necessary to tow it by means of a line.

CHAPTER XVII.

Voyage up the Orinoco.

Ascent of the Orinoco—Port of Encaramada—Traditions of a Universal Deluge—Gathering of Turtles' Eggs—Two Species described—Mode of collecting the Eggs and of manufacturing the Oil—Probable Number of these Animals on the Orinoco—Decorations of the Indians—Encampment of Pararuma—Height of the Inundations of the Orinoco—Rapids of Tabage.

LEAVING the Rio Apure the travellers entered the Orinoco, and presently found themselves in a country of an entirely different aspect. As far as the eye could reach there lay before them a sheet of water, the waves of which, from the conflict of the breeze and the current, rose to the height of several feet. The long files of herons, flamingoes, and spoonbills, which were observed on the Apure, had disappeared; and all that supplied the place of those multitudes of animated beings by whom they had been lately accompanied, was here and there a crocodile swimming in the agitated stream. The horizon was bounded by a girdle of forests, separated from the river by a broad beach, the bare and parched surface of which refracted the solar rays into the semblance of pools.

The wind was favourable for sailing up the Orinoco; but the short broken waves at the junction of the two rivers were exceedingly disagreeable. They passed the Punta Curiquima, a granitic promontory, between which and the mouth of the

Apure, the breadth of the stream was ascertained to be 4063 yards, and in the rainy season it extends to 11,760. The temperature of the water was in the middle of the current 82·9°, and near the shores, 84·6°. They first went up toward the southwest as far as the shore of the Guaricoto Indians on the left bank, and then toward the south. The mountains of Encaramada, forming a continued chain from west to east, seemed to rise from the water as distant land rises on the horizon at sea. The beach was composed of clay intermixed with scales of mica, deposited in very thin strata. At the port of Encaramada, where they stopped for some time, they met with a Carib cacique going up the river in his canoe to gather turtles' eggs. He was armed with a bow and arrows, as were his attendants, and, like them, he was naked and painted red. These Indians were tall and athletic, and, with their hair cut straight across the forehead, their eyebrows painted black, and their gloomy but animated countenances, had a singular appearance. The travellers were surprised to find that the anterior portion of the cranium is not so depressed as those of the Caribs are usually represented to be. The women carried their infants on their backs. The shore is here formed by a rock forty or fifty feet high, composed of blocks of granite piled upon each other; the surface of which was of a dark-gray colour, although the interior was reddish-white. The night was passed in a creek opposite the mouth of the Rio Cabullare. The evening was beautiful, with moonlight; but towards twelve the north-east wind blew so violently that they became apprehensive for the safety of their canoe.

On the 6th, continuing to ascend, they saw the southern side of the mountains of Encaramada, which stretch along the right bank of the river, and are inhabited by Indians of a gentle character, and addicted to agriculture. There is a tradition here, and elsewhere on the Orinoco, among the natives, " That at the time of the Great Waters, when their fathers were obliged to betake themselves to their canoes in order to escape the general inundation, the waves of the sea beat upon the rocks of Encaramada." When the Tamanacs are asked how the human race survived this great deluge, they say, " That a man and a woman saved themselves upon a high mountain called Tamanacu, situated on the bank of the Aseveru, and that, throwing behind them, over their heads, the fruits of the Mauritia palm, they saw arising from the nuts of these fruits the men and women who repeopled the earth." Thus, among the natives of America, a fable similar to that of Pyrrha and Deucalion commemorates the grand catastrophe of a general inundation. Humboldt, in reference to the same event, mentions that hieroglyphic figures are often found along the Orinoco sculptured on rocks now inaccessible but by scaffolding, and that the natives, when asked how these figures could have been made, answer with a smile, as relating a fact of which a stranger alone could be ignorant, " That at the period of the Great Waters their fathers went to that height in boats."

" These ancient traditions of the human race," says Humboldt, " which we find dispersed over the surface of the globe, like the fragments of a vast shipwreck, are of the greatest interest in the philosophical study of our species. Like certain families of plants, which, notwithstanding the diversity of climates and the influence of heights, retain the impress of a common type, the traditions respecting the primitive state of the globe present among all nations a resemblance that fills us with astonishment; so many different languages belonging to branches which appear to have no connexion with each other, transmit the same facts to us. The substance of the traditions respecting the destroyed races and the renovation of nature is every where almost the same, although each nation gives it a local colouring. In the great continents, as in the smallest islands of the Pacific Ocean, it is always on the highest and nearest mountain that the remains of the human race were saved; and this event appears so much the more recent the more uncultivated the nations are, and the shorter the period since they have begun to acquire a knowledge of themselves. When we attentively examine the Mexican monuments anterior to the discovery of America,—penetrate into the forests of the Orinoco, and become aware of the smallness of the European establishments, their solitude, and the state of the tribes which retain their independence,—we cannot allow ourselves to attribute the agreement of these accounts to the influence of missionaries and to that of Christianity upon national traditions. Nor is it more probable that the sight of marine bodies, found on the summits of mountains, presented to the tribes of the Orinoco the idea of those great inundations which for some time extinguished the germs of organic life upon the globe.—The country which extends from the right bank of the Orinoco to the

Casiquiare and the Rio Negro consists of primitive rocks. I saw there a small deposite of sandstone or conglomerate, but no secondary limestone, and no trace of petrifactions."

At eleven in the morning the travellers landed on an island celebrated for the turtle fishery, or the "harvest of eggs," which takes place annually. Here they found encamped more than 300 Indians of different races, each tribe, distinguished by its peculiar mode of painting, keeping separate from the rest, together with a few white men who had come to purchase egg-oil from them. The missionary of Uruana, whose presence was necessary to procure a supply for the lamp of the church and keep the natives in order, received the strangers with kindness, and made the tour of the island with them; showing them, by means of a pole which he thrust into the sand, the extent of the stratum of eggs, that had been deposited wherever there were no eminences. The Indians asserted, that in coming up the Orinoco, from its mouth to the junction of the Apure, there is no place where eggs can be collected in abundance; and the only three spots where the turtles assemble annually in great numbers are situated between the mouth of the Apure and the great cataracts. These animals do not seem to pass beyond the falls, the species found above Atures and Maypures being different.

The arrau or tortuga, which deposites the eggs that are so much valued on the Lower Orinoco, is a large fresh-water tortoise, with webbed feet, a very flat head, a deep groove between the eyes, and an upper shell composed of five central, eight lateral, and twenty-four marginal scutella or plates. The colour is dark-gray above and orange beneath.

When of full size it weighs from forty to fifty pounds. The eggs are much larger than those of a pigeon, and are covered with a calcareous crust.

The terekay, the species which occurs above the cataracts, is much smaller. It has the same number of dorsal plates, but the colour is olive green, with two spots of red mixed with yellow on the top of the head, and a prickly appendage under the chin. The eggs have an agreeable taste, and are much sought after, but are not deposited in masses like those of the tortuga. This variety is found below the cataracts as well as in the Apure, the Urituco, the Guarico, and the small rivers of the Llanos of Caraccas.

The period at which the arrau deposites its eggs is when the river is lowest. About the beginning of February these creatures issue from the water and warm themselves on the beach, remaining there a great part of the day. Early in the month of March they assemble on the islands where they breed, when thousands are to be seen ranged in files along the shores. The Indians place sentinels at certain distances, to prevent them from being disturbed, and the people who pass in boats are told to keep in the middle of the river. The laying of the eggs begins soon after sunset, and is continued throughout the night. The animal digs a hole three feet in diameter and two in breadth with its hind feet, which are very long and furnished with crooked claws. So pressing is the desire which it feels to get rid of its burden, that great confusion prevails, and an immense number of eggs is broken. Some of the tortoises are surprised by day before they have finished the operation, and, becoming insensible to danger, continue to work with the greatest diligence even in the presence of the fishers.

The Indians assemble about the beginning of April, and commence operations under the direction of the missionaries, who divide the egg-ground into portions. The leading person among them first examines, by means of a long pole or cane, how far the bed extends, and then allots the shares. The natives remove the earth with their hands, gather up the eggs, and carry them in baskets to the camp, where they throw them into long wooden troughs filled with water. They are next broken and stirred, and remain exposed to the sun, until the yolk, which swims at the surface, has time to inspissate, when it is taken off and boiled. The oil thus obtained is limpid and destitute of smell, and is used for lamps as well as for cooking. The shores of the missions of Uruana furnish 1000 botijas or jars annually, and the three stations jointly may be supposed to furnish 5000. It requires 5000 eggs to fill a jar; and if we estimate at 100 or 116 the number which one tortoise produces, and allow one-third to be broken at the time of laying, we may presume that 330,000 of these animals assemble every year, and lay 33,000,000 of eggs. This calculation, however, is much below the truth. Many of them lay only 60 or 70; great numbers of them again are devoured by jaguars; the Indians take away a considerable quantity to eat them dried in the sun, and break nearly as many while gathering them; and, besides, the proportion that is hatched is such, that Humboldt saw the whole shore near the encampment of Uruana swarming with young ones. Moreover, all the arraus do not assemble on the three shores of the encampments, but many lay elsewhere. The number which annually deposite their eggs on the shores of the Lower Orinoco may, therefore, be estimated at little short of a million. The travellers were shown the shells of large turtles which had been emptied by the jaguars. These animals surprise them on the sand, and turn them on their back in order to devour them at their ease; they dig up the eggs also; and, together with the gallinazo vulture and the herons, destroy thousands of their brood.

After procuring some fresh provision, and taking leave of the missionary, they set sail in the afternoon. The wind blew in squalls, and after they had entered the mountainous part of the country, they found the canoe not very safe when under sail; but the master was desirous of showing off to the Indians, and in going close upon the wind almost upset his vessel, which filled with water, and nearly foundered. In the evening they landed on a barren island, where they supped under a beautiful moonlight, with turtle-shells for seats, and indulged their imagination with the picture of a shipwrecked man, wandering on the desert shores of the Orinoco amid rivers full of crocodiles and caribe fishes. The night was intensely hot, and not finding trees on which to sling their hammocks, they slept on skins spread on the ground. To their surprise the jaguars swam to the island, although they had kindled fires to prevent them; but these animals did not venture to attack them.

On the 7th they passed the mouth of the Rio Arauca, which is frequented by immense numbers of birds. They also saw the mission of Uruana, at the foot of a mountain composed of detached blocks of granite, in the caverns formed by which hieroglyphic figures are sculptured. Measuring the breadth of the Orinoco here, they found it, at a distance of 670 miles from the mouth, to be 5700 yards, or nearly three miles. The temperature of the water at its surface was 82°. As the strength of the current increased the progress of the boat became much slower, while at one time the woods deprived them of the wind, and at another a violent gust descended from the mountain-passes. Opposite the lake of Capanaparo, which communicates with the river, the number of crocodiles was increased. The Indians asserted that they came in troops to the water from the savannahs, where they lie buried in the solid mud until the first showers awaken them. Humboldt remarks, that the dry season of the torrid zone corresponds to the winter of the temperate regions of the globe; and that while the alligators of North America become torpid through excess of cold, the crocodiles of the Llanos are reduced to the same state through deficiency of moisture.

They now entered the passage of the Baraguan, where the Orinoco is hemmed in by precipices of granite, forming part of a range of mountains through which it has found or forced a channel. Like all the other granitic hills which they observed on this river, they were formed of enormous cubical masses piled upon each other. Landing in the middle of the strait, they found the breadth of the stream to be 1895 yards. They looked in vain for plants in the fissures of the rocks; but the stones were covered with multitudes of lizards. There was not a breath of wind, and the heat was so intense that the thermometer placed against the rock rose to 122·4°. " How vivid," says Humboldt, " is the impression which the noontide quiet of nature produces in these burning climates! The beasts of the forest retire to the thickets, and the birds conceal themselves among the foliage or in the crevices of rocks. Yet amid this apparent silence, should one listen attentively, he hears a stifled sound, a continued murmur, a hum of insects, that fill the lower strata of the air. Nothing is more adapted to excite in man a sentiment of the extent and power of organic life. Myriads of insects crawl on the ground, and flutter round the plants scorched by the heat of the sun. A confused noise issues from every bush, from the decayed trunks of the trees, the fissures of the rocks, and from the ground, which is undermined by lizards, millipedes, and blindworms. It is a voice proclaiming to us that all nature breathes, that under a thousand different forms life is diffused in the cracked and dusty soil, as in the bosom of the waters, and in the air that circulates around us." The water of the river was very disagreeable here, as it had a musky smell and a sweetish taste. In some parts it was pretty good; but in others it seemed loaded with gelatinous matter, which the natives attribute to putrified crocodiles.

After sleeping at the foot of an eminence they continued their voyage, and passed the mouths of several rivers; and on the 9th arrived, early in the morning, at the beach of Pararuma, where they found an encampment of Indians, who had assembled to search the sands for turtles' eggs. The pilot, who had brought them from San Fernando de Apure, would not undertake to accompany them farther; but they procured a boat from one of the missionaries who had come to the egg-harvest.

This assemblage or encampment afforded to the travellers an interesting subject of study. "How difficult," says Humboldt, "to recognise in this infancy of society, this collection of dull, taciturn, and unimpassioned Indians, the original character of our species! Human nature is not seen here arrayed in that gentle simplicity of which poets in every language have drawn such enchanting pictures. The savage of the Orinoco appeared to us as hideous as the savage of the Mississippi, described by the philosophical traveller who best knew how to paint man in the various regions of the globe. One would fain persuade himself that these natives of the soil, crouched near the fire, or seated on large shells of turtles, their bodies covered with earth and grease, and their eyes stupidly fixed for whole hours on the drink which they are preparing, far from being the original type of our species, are a degenerated race, the feeble remains of nations which, after being long scattered in the forests, have been again immersed in barbarism."

Red paint is the ordinary decoration of these tribes. The most common kind is obtained from the seeds of the *Bixa orellana,* and is called anotto, achote, or roucou. Another much more expensive species is extracted from the leaves of *Bignonia chica.* Both these are red; but a black ingredient is obtained from the *Genipa Americana,* and is called caruto. These pigments are mixed with turtle-oil or grease, and are variously applied according to national or individual taste. The Caribs and Otomacs colour only the head and hair, while the Salivas smear the whole body; but there prevails in general as great a diversity in the mode of staining as is found in Europe in respect to dress; and at Pararuma the travellers saw some Indians painted with a blue jacket and black buttons. Women advanced in years are fonder of being thus ornamented than the younger ladies; and so expensive is this mode of decoration, that an industrious man can hardly gain enough by the labour of a fortnight to adorn himself with chica, of which the missionaries make an article of traffic. After all, the paintings that cost so much are liable to be effaced by a heavy shower; although the caruto long resists the action of water, as the travellers found by disagreeable experience; for having one day in sport marked their faces with spots and strokes of it, it was not entirely removed till after a long period. It has been supposed that this usage prevents the Indians from being stung by insects; but this was found to be incorrect. The preference given by the American tribes to the red colour, Humboldt supposes to be owing to the tendency which nations feel to attribute the idea of beauty to whatever characterizes their national complexion.

The encampment of Pararuma also afforded the travellers an opportunity of examining several animals they had not before seen alive, and which the Indians brought to exchange with the missionaries for fish-hooks and other necessaries. Among these specimens were gallitoes, or rock-manakins, monkeys of different species, of which the titi or *Simia sciurea* seems to have been a special favourite with Humboldt. He mentions a very interesting fact illustrative of the sagacity of this creature. One which he had purchased of the natives distinguished the different plates of a work on natural history so well, that when an engraving which contained zoological representations was placed before it, it rapidly advanced its little hand to catch a grashopper or a wasp; which was the more remarkable as the figures were not coloured. Humboldt observes, that he never heard of any the most perfect picture of hares or deer producing the least effect upon a hound, and doubts if there be a well-ascertained example of a dog having recognised a full-length portrait of its master.

The canoe which they had procured was forty-two feet long and three broad. The missionary of Atures and Maypures had offered to accompany them as far as the frontiers of Brazil, and made preparations for the voyage. Two Indians who were to form part of the crew were chained during the night to prevent their escape; and on the morning of the 10th the company set out. The vessel was found to be extremely incommodious. To gain something in breadth a kind of frame had been extended over the gunwale in the hinder part of it; but the roof of leaves which covered it was so low, that the travellers were obliged to lie down, or sit nearly double, while in rainy weather the feet were liable to be wetted. The natives, seated two and two, were furnished with paddles three feet long, and rowed with surprising uniformity to the cadence of a monotonous and melancholy song. Small cages containing birds and monkeys were suspended to the shed, and the dried plants and instruments were placed beneath it. To their numerous inconveniences was added the continual torment of the mosquitoes, which they were unable by any means to alleviate. Every night, when they established their watch, the collection of animals and instruments occupied the centre, around which were placed first their own hammocks, and then those of the Indians, while fires were lighted to intimidate the jaguars. At sunrise the monkeys in the cages answered the cries of those in the forests, affording an affecting display of sympathy between the captive and the free.

Above the deserted mission of **Pararuma** the river is full of islands, and divides into several branches. Its total breadth is about 6395 yards. The country becomes more wooded. A granitic prism, terminated by a flat surface covered with a tuft of trees, rises to the height of 213 feet in the midst of the forest. Farther on the river narrows; and upon the east is an eminence, on which the Jesuits formerly maintained a garrison for protecting the missions against the inroads of the Caribs, and for extending what, in the Spanish colonies, was called the conquest of souls, which of course was effected through the conquest of bodies. The soldiers made incursions into the territories of the independent Indians, killed all who offered resistance, burned their huts, destroyed the plantations, and made prisoners of the old men, women, and children, who were afterwards divided among their establishments. The river again contracted, and rapids began to make their appearance, the shores becoming sinuous and precipitous. In a bay between two promontories of granite, they landed at what is called the Port of Carichana, and proceeded to the mission of that name, situated at the distance of two miles and a half from the bank, where they were hospitably received at the priest's house. The Christian converts at this station were Salivas, a social and mild people, having a great taste for music.

Among these Indians they found a white woman, the sister of a Jesuit of New Grenada, and experienced great pleasure in conversing with her without the aid of a third person. In every mission, says Humboldt, there are at least two interpreters, for the purpose of communicating between the monks and the catechumens, the former seldom studying the language of the latter. They are natives, somewhat less stupid than the rest, but ill adapted for their office. They always attended the travellers in their excursions; but little more could be got from them than a mere affirmation or negation. Sometimes, in attempting to hold intercourse with the Indians, he preferred the language of signs,—a method which he recommends to travellers, as the variety of languages spoken on the Meta, Orinoco, Casiquiare, and Rio Negro, is so great, that no one could ever make himself understood in them all.

The scenery around the mission of Carichana appeared delightful. The village was situated on a grassy plain, bounded by mountains. Banks of rock, often more than 850 feet in circumference, scarcely elevated a few inches above the savannahs, and nearly destitute of vegetation, give a peculiar character to the country. On these stony flats they eagerly observed the rising vegetation in the different stages of its development: Lichens cleaving the rock and collected into crusts; a few succulent plants growing among little portions of quartz-sand; and tufts of evergreen shrubs springing up in the black mould deposited in the hollows. At the distance of eight or ten miles from the religious house they found a rich and diversified assemblage of plants, among which M. Bonpland obtained numerous new species. Here grew the *Dipterix odorata,* which furnishes excellent timber, and of which the fruit is known in Europe by the name of tonkay or tongo bean.

In a narrow part of the river the marks of the great inundations were 45 feet above the surface; but at various places black bands and erosions are seen, 106 or even 138 feet above the present highest increase of the waters. " Is this river, then," says Humboldt, " the Orinoco, which appears to us so imposing and majestic, merely the feeble remnant of those immense currents of fresh water which, swelled by Alpine snows or by more abundant rains, every where shaded by dense forests and destitute of those beaches that favour evaporation, formerly traversed the regions to the east of the Andes, like arms of inland seas? What must then have been the state of those low countries of Guiana, which now experience the effects of annual inundations? What a prodigious number of crocodiles, lamantines, and boas, must have inhabited these vast regions, alternately converted into pools of stagnant water and arid plains! The more peaceful world in which we live has succeeded to a tumultuous world. Bones of mastodons and real American elephants are found dispersed over the platforms of the Andes. The megatherium inhabited the plains of Uruguay. By digging the earth more deeply in high valleys, which at the present day are unable to nourish palms or tree-ferns, we discover strata of coal containing gigantic remains of monocotyledonous plants. There was therefore a remote period, when the tribes of vegetables were differently distributed; when the animals were larger, the rivers wider and deeper. There stop the monuments of nature which we can consult. We are ignorant if the human race, which at the time of the discovery of America scarcely presented a few feeble tribes to the east of the Cordilleras, had yet descended into the plains, or if the ancient tradition of the Great Waters, which we find among all the races of the Orinoco, Erevato, and Caura, belong to other climates, whence it had been transferred to this part of the new continent."

On the 11th they left Carichana at two in the afternoon, and found the river more and more encumbered by blocks of granite. At the large rock known by the name of Piedra del Tigre, the depth is so great that no bottom can be found with a line of 140 feet. Towards evening they encountered a thunder-storm, which for a time drove away the mosquitoes that had tormented them during the day. At the cataract of Cariven the current was so rapid that they had great difficulty in landing; but at length two Saliva Indians swam to the shore, and drew the canoe to the side with a rope. The thunder continued a part of the night, and the river increased considerably. The granitic rock on which they slept, is one of those from which travellers on the Orinoco have heard subterranean sounds, resembling those of an organ, emitted about sunrise. Humboldt supposes that these must be produced by the passage of rarefied air through the fissures, and seems to think, that the impulse of the fluid against the elastic scales of mica which intercept the crevices may contribute to modify their expression.*

On the 12th they set off at four in the morning. The Indians rowed twelve hours and a half without intermission, during which time they took no other nourishment than cassava and plantains. The bed of the river, to the length of 1280 yards, was full of granite rocks, the channels between which were often very narrow, insomuch, that the canoe was sometimes jammed in between two blocks. When the current was too strong the sailors leapt out, and warped the boat along. The rocks were of all dimensions, rounded, very dark, glossy like lead, and destitute of vegetation. No crocodiles were seen in these rapids. The left bank of the Orinoco, from Cabruto to the mouth of the Rio Serianico, a distance of nearly two degrees of latitude, is entirely

---

* Many examples of mysterious sounds produced under similar circumstances are on record. In the autumn of 1828, a recent traveller crossing the Pyrenees, when in a wild pass with the Maladetta mountain opposite, heard "a dull, low, moaning, Æolian sound, which alone broke upon the deathly silence, evidently proceeding from the body of this mighty mass." The air was perfectly calm, and clear to an extraordinary degree; no waterfall could be seen even with the aid of a telescope, and no cause could be assigned for the phenomenon, unless the sun's rays, "at that moment impinging in all their glory on every point and peak of the snowy heights," had some share "in vibrating these mountain-chords."—*N. M. Mag.* xxx. 341.——The granite statue of Memnon is well known to have emitted sounds when the morning beams darted upon it; and MM. Jomard, Jollois, and Devilliers, heard a noise resembling that of the breaking of a string, which proceeded at sunrise from a monument of granite situated near the centre of the spot on which stands the palace of Carnac. Singular sounds have been heard from the interior of a mountain near Tor, in Arabia Petræa. They are familiar to the natives, who ascribe them to a convent of monks miraculously preserved under ground, and were heard by M. Seetzen and Mr Gray, the only European travellers who have visited the place. For an account of these curious phenomena, the reader may be referred to Dr Brewster's Letters on Natural Magic, forming No. XXXIII. of the Family Library.

uninhabited; but to the westward of these rapids an enterprising individual, Don Felix Relinchon, had formed a village of Jaruro and Otomac Indians. At nine in the morning they arrived at the mouth of the Meta, which, next to the Guaviare, is the largest river that joins the Orinoco. At the union of these streams the scenery is of a very impressive character. Solitary peaks rise on the eastern side, appearing in the distance like ruined castles, while vast sandy shores intervene between the bank and the forests. They passed two hours on a large rock in the middle of the Orinoco, upon which Humboldt succeeded in fixing his instruments, and in determining the longitude of the embouchure of the Meta; a river which will one day be of great political importance to the inhabitants of Guiana and Venezuela, as it is navigable to the foot of the Andes of New Grenada. Above this point the current was comparatively free from shoals; and in the evening they reached the Rapids of Tabaje. As the Indians would not venture to pass them they were obliged to land and repose on a craggy platform having a slope of more than eighteen degrees, and having its crevices filled with bats. The cries of the jaguar were heard very near during the whole night; the sky was of a tremendous blackness; and the hoarse noise of the rapids blended with the thunder which rolled at a distance amongst the woods.

Early in the morning they cleared the rapids, and disembarked at the new mission of San Borja, where they found six houses inhabited by uncatechised Guahiboes, who differed in nothing from the wild natives. The faces of the young girls were marked with black spots. This people had not painted their bodies, and several of them had beards, of which they seemed proud, taking the travellers by the chin, and showing by signs that they were like themselves. In continuing to ascend the river they found the heat less intense, the temperature during the day being 79° or 80°, and at night about 75°; but the torment of the mosquitoes increased. The crocodiles which they saw were all of the extraordinary size of 24 or 25 feet.

The night was spent on the beach; but the sufferings inflicted by the flies induced the travellers to start at five in the morning. On the island of Guachaco, where they stopped to breakfast, they found the granite covered by a sandstone or conglomerate, containing fragments of quartz and felspar cemented by indurated clay, and exhibiting small veins of brown iron-ore. Passing the mouth of the Rio Parueni, they slept on the island of Panumana, which they found rich in plants, and where they again observed the low shelves of rock partially coated with the vegetation which they had admired at Carichana.

CHAPTER XVIII.

Voyage up the Orinoco continued.

Mission of Atures—Epidemic Fevers—Black Crust of Granitic Rocks—Causes of Depopulation of the Missions—Falls of Apures—Scenery—Anecdote of a Jaguar—Domestic Animals—Wild Man of the Woods—Mosquitoes and other poisonous Insects—Mission and Cataracts of Maypures—Scenery—Inhabitants—Spice-trees—San Fernando de Atabapo—San Baltasar—The Mother's Rock—Vegetation—Dolphins—San Antonio de Javita—Indians—Elastic Gum—Serpents—Portage of the Pimichin—Arrival at the Rio Negro, a Branch of the Amazon—Ascent of the Casiquiare.

Leaving the island of Panumana at an early hour the navigators continued to ascend the Orinoco, the scenery on which became more interesting the nearer they approached the great cataracts. The sky was in part obscured, and lightnings flashed among the dense clouds; but no thunder was heard. On the western bank of the river they perceived the fires of an encampment of Guahiboes, to intimidate whom some shots were discharged by the direction of the missionary. In the evening they arrived at the foot of the great fall, and passed the night at the mission of Atures in its neighbourhood. The flat savannah which surrounds the village seemed to Humboldt to have formerly been the bed of the Orinoco.

This station was found to be in a deplorable state, the Indians having gradually deserted it until only forty-seven remained. At its foundation in 1748 several tribes had been assembled, which subsequently dispersed, and their places were supplied by the Guahiboes, who belong to the lowest grade of uncivilized society, and a few families of Macoes. The epidemic fevers, which prevail here at the commencement of the rainy season, contributed greatly to the decay of the establishment. This distemper is ascribed to the violent heats, excessive humidity of the air, bad food, and, as the natives believe, to the noxious exhalations that rise from the bare rocks of the rapids. This last is a curious circumstance, and, as Humboldt remarks, is the more worthy of attention on account of its being connected with a fact that has been observed in several parts of the world, although it has not yet been sufficiently explained.

Among the cataracts and falls of the Orinoco, the granite rocks, wherever they are periodically submersed, become smooth and seem as if coated with black lead. The crust is only 0·3 of a line in thickness, and occurs chiefly on the quartzy parts of the stone, which is coarse grained, and contains solitary crystals of hornblende. The same appearance is presented at the cataracts of Syene as well as those of the Congo. This black deposite, according to Mr Children's analysis, consists of oxide of iron and manganese, to which some experiments of Humboldt induced him to add carbon and supercarburetted iron. The phenomenon has hitherto been observed only in the torrid zone, in rivers that overflow periodically and are bounded by primitive rocks, and is supposed by our author to arise from the precipitation of substances chemically dissolved in the water, and not from an efflorescence of matters contained in the rocks themselves. The Indians and missionaries assert, that the exhalations from these rocks are unwholesome, and consider it dangerous to sleep on granite near the river; and our travellers, without entirely crediting this assertion, usually took care to avoid the black rocks at night. But the danger of reposing on them, Humboldt thinks, may rather be owing to the very great degree of warmth they retain during the night, which was found to be 85·5°, while that of the air was 78·8°. In the day their temperature was 118·4°, and the heat which they emitted was stifling.

Among the causes of the depopulation of the missions, Humboldt mentions the general insalubrity of the climate, bad nourishment, want of proper treatment in the diseases of children, and the practice of preventing pregnancy by the use of deleterious herbs. Among the savages of Guiana, when twins are produced one is always destroyed, from the idea that to bring more than one at a time into the world is to resemble rats, opossums, and the vilest animals, and that two children born at once cannot belong to the same father. When any physical deformity occurs in an infant, the father puts it to death, and those of a feeble constitution sometimes undergo the same fate, because the care which they require is disagreeable. "Such," says Humboldt, " is the simplicity of manners,—the boasted happiness of man, in the state of nature! He kills his son to escape the ridicule of having twins, or to avoid travelling more slowly,—in fact, to avoid a little inconvenience."

The two great cataracts of the Orinoco are formed by the passage of the river across a chain of granitic mountains, constituting part of the Parime range. By the natives they are called Mapara and Quittuna; but the missionaries have denominated them the falls of Atures and Maypures, after the first tribes which they assembled in the nearest villages. They are only 41 miles distant from each other, and are not more than 345 miles west of the cordilleras of New Grenada. They divide the Christian establishments of Spanish Guiana into two unequal parts; those situated between the lower cataract, or that of Apures, being called the missions of the Lower Orinoco, and those between the upper cataract and the mountains of Duida, being called the missions of the Upper Orinoco. The length of the lower section, including its sinuosities, is 897 miles, while that of the upper is 576 miles. The navigation of the river extends from its mouth to the point where it meets the Anaveni near the lower cataract, although in the upper part of this division there are rapids which can be passed only in small boats. The principal danger, however, is that which arises from natural rafts, consisting of trees interwoven with lianas, and covered with aquatic plants carried down by the current. The cataracts are formed by bars stretching across the bed of the river, which forces its way through a break in the mountains; but beyond this rugged pass the course is again open for a length of more than 576 miles.

The scenery in the vicinity of the lower fall is described as exceedingly beautiful. To the west of Atures, a pyramidal mountain, the Peak of Uniana, rises from a plain to the height of nearly 3200 feet. The savannahs, which are covered with grasses and slender plants, though never inundated by the river, present a surprising luxuriance and diversity of vegetation. Piles of granitic blocks rise here and there, and at the margins of the plains occur deep valleys and ravines, the humid soil of which is covered with arums, heliconias, and lianas. The shelves of primitive rocks, scarcely elevated above the plain, are partially coated with lichens and mosses, together with succulent plants, and tufts of evergreen shrubs with shining leaves. On all sides the horizon is bounded by mountains, overgrown with forests of laurels, among which clusters of palms rise to the height of more than a hundred feet, their slender stems supporting tufts of feathery foliage. To the east of Atures other mountains appear, the ridge of which is composed of pointed cliffs, rising like huge pillars above the trees. When these columnar masses are situated near the Orinoco, flamingoes, herons, and other wading birds, perch on their summits, and look like sentinels. In the vicinity of the cataracts, the moisture which is diffused in the air produces a perpetual verdure, and wherever soil has accumulated on the plains, it is occupied by the beautiful shrubs of the mountains.

The rainy season had scarcely commenced, yet the vegetation displayed all the vigour and brilliancy which, on the coast, it assumes only towards the end of the rains. The old trunks were decorated with orchideæ, bannisterias, bignonias, arums, and other parasitic plants. Mimosas, figs, and laurels, were the prevailing trees in the woody spots; and in the vicinity of the cataract were groups of heliconias, bamboos, and palms.

Along a space of more than five miles, the bed of

the Orinoco is traversed by numerous dikes of rock, forming natural dams, filled with islands of every form, some rocky and precipitous, while others resemble shoals. By these the river is broken up into torrents, which are ever dashing their spray against the rocks. They are all furnished with sylvan vegetation, and resemble a mass of palm-trees rising amidst the foam of the waters. The current is divided into a multitude of rapids, each endeavouring to force a passage through the narrows, and is every where engulfed in caverns, in one of which the travellers heard the water rolling at once over their heads and beneath their feet.

Notwithstanding the formidable aspect of this long succession of falls, the Indians pass many of them in their canoes. When ascending they swim on before, and after repeated efforts succeed in fixing a rope to a point of rock, and thus draw the canoe up the rapid. Sometimes it fills with water, and is not unfrequently dashed to pieces against the shelves, upon which the sailors again swim, though not without difficulty, through the whirlpools to the nearest island. When the bars are very high the vessels are taken ashore, and drawn upon rollers, made of the branches of trees, to a place where the river again becomes navigable. During the flood, however, this operation is seldom necessary.

Although the rapids of the Orinoco form a long series of falls, the noise of which is heard at the distance of more than three miles, yet the rocks were found by Humboldt not to have a greater height than thirty feet perpendicular. He thinks it probable that a considerable part of the water is lost by passing into subterranean cavities, independently of that which disappears by being dispersed in the atmosphere. Numberless holes and sinuosities are formed in the crevices by the friction of the sand and quartz pebbles; but he does not consider that any great change is effected in the general form of the cataracts by the action of the water, the granite being too hard to be worn away to a great extent. The Indians assert that the stony barriers preserve the same aspect; but that the partial torrents into which the river divides itself are changed in their direction, and carry sometimes more sometimes less water towards one or other bank.

When the rush of the cataracts is heard in the plain that surrounds the mission of Atures, one imagines he is near a coast skirted by reefs and breakers. The noise is thrice as loud by night as by day. This circumstance had struck the padre and the Indians, and Humboldt attributes it to the cessation of the sun's action, which is productive of numberless currents and undulations of the air, impeding the progress of sound by presenting spaces of different density.

The jaguars, which abound every where on the Orinoco, are so numerous here that they come into the village, and devour the pigs of the poor Indians. The missionary related a striking instance of the familiarity of these animals:—" Two Indian children, a boy and girl eight or nine years of age, were sitting among the grass near the village of Atures, in the midst of a savannah. It was two in the afternoon when a jaguar issued from the forest and approached the children, gamboling around them; sometimes concealing itself among the long grass, and again springing forward, with his back curved and his head lowered, as is usual with our cats. The little boy was unaware of the danger in which he was placed, and became sensible of it only when the jaguar struck him on the head with one of his paws. The blows thus inflicted were at first slight, but gradually became ruder. The claws of the jaguar wounded the child, and blood flowed with violence. The little girl then took up a branch of a tree and struck the animal, which fled before her. The Indians hearing the cries of the children, ran up and saw the jaguar, which bounded off without showing any disposition to defend itself." "What," asks Humboldt, "meant this fit of playfulness in an animal which, although not difficult to be tamed in our menageries, is always so ferocious and cruel in the state of freedom? If we choose to admit that, being sure of its prey, it played with the young Indian as the domestic cat plays with a bird, the wings of which have been clipped, how can we account for the forbearance of a large jaguar when pursued by a little girl? If the jaguar was not pressed by hunger, why should it have gone up to the children? There are mysteries in the affections and hatreds of animals. We have seen lions kill three or four dogs which were put into their cage, and instantly caress another which had the courage to seize the royal beast by the mane. Man is ignorant of the sources of these instincts. It would seem that weakness inspires more interest the more confiding it is."

The cattle introduced by the Jesuits had entirely disappeared; but the Indians rear the common pig and another kind peculiar to America, and known in Europe by the name of pecari. A third species of hog, the Apida, which is of a dark-brown colour, wanders in large herds composed of several hundreds. M. Bonpland, when upon a botanical excursion, saw a drove of these animals pass near him. It marched in a close body; the males before, and each sow accompanied by her young. The natives kill them with small lances tied to cords. At the mission they saw a monkey of a new species, which had been brought up in captivity, and which every day seized a pig in the court-yard, and remained upon it from morning to night, in all its wanderings in the savannahs. Here, for the first time, they heard of the hairy man of the woods, a large animal of the ape kind, which, according to report, carries off women, builds huts, and sometimes eats human flesh. Father Gili gravely relates the history of a lady of San Carlos, who passed several years with one, which she left only because she and the children she had to him were tired of living far from the church and the sacraments. In all his travels in America, Humboldt found no traces of a large anthropomorphous monkey, although in several places, very distant from each other, he heard similar accounts of it.

Flies of various kinds unceasingly tormented the travellers; mosquitoes and simulia by day, and zancudoes by night. The missionary, observing that the insects were more abundant in the lowest stratum of the atmosphere, had constructed near the church a small apartment supported upon palm-trunks, to which they retired in the evening to dry their plants and write their journals.* At May-

---

* A similar expedient was tried by a British officer who had joined the insurgents under Bolivar, in 1818. "These insects," (the pures the Indians leave the village at night, and sleep on the little islands in the midst of the cataracts, where the insects are less numerous. Humboldt gives an elaborate account of these creatures, of which, however, the most interesting particulars alone can be here extracted. In the missions of the Orinoco, when two persons meet in the morning, the first questions are,—" How did you find the zancudoes during the night? How are we to-day for the mosquitoes?" The plague of these animals, however, is not so general in the torrid zone as is commonly believed. On the table-lands that have an elevation of more than 2558 feet, and in very dry plains at a distance from rivers, they are not more numerous than in Europe; but along the valleys, as well as in moist places on the coast, they continually harass the traveller; the lower stratum of air, to the height of fifteen or twenty feet, being filled with a cloud of venomous insects. It is a remarkable circumstance that on the streams, the water of which is of a yellowish-brown colour, the tipulary flies do not make their appearance. Not less astonishing is the fact, that the different kinds do not associate together; but that

---

mosquitoes), says he, " do not rise high in the air, but are generated and remain near the wet banks of the river. I found a tree in the neighbourhood, which I ascended nearly to its top with a cord. This I attached firmly to the branches, and then fixed it round me, so that I could not fall, but sit with safety, although not with much comfort. It was, however, with me here as with many in various situations in life—I could estimate the nature and extent of my pleasures and my difficulties merely by comparison; and, certainly, although the being tied to the top of a tree as a sleeping-place was not very agreeable, it was far preferable to being among swarms of hungry mosquitoes where I had previously lodged. I enjoyed several hours' sleep and awoke considerably refreshed."—*Robinson's Journal of an Expedition up the Orinoco and Arauca.*

at certain hours of the day, distinct species, as the missionaries say, mount guard. From half after six in the morning till five in the afternoon the air is filled with mosquitoes, which are of the genus *Simulium*, and resemble a common fly. An hour before sunset small gnats, called tempraneroes, succeeded them, to disappear between six and seven; after which zancudoes, a species of gnat with very long legs, come abroad and continue until near sunrise, when the former again take their turn. Persons born in the country, whether whites, mulattoes, negroes, or Indians, all suffer from the sting of these insects, although not so severely as recently-arrived Europeans.

The travellers, after remaining two days in the vicinity of the cataract of Atures, proceeded on the 17th to rejoin their canoe, already conducted by eight Indians of the mission through the rapids, and reached it about eleven in the morning, accompanied by Father Zea, who had procured a small stock of provisions, consisting of plantains, cassava, and fowls. The river was now free from shoals; and after a few hours they passed the rapids of Garcita, and perceived numerous small holes, at an elevation of more than 190 feet above the level of the current, which appeared to have been caused by the erosion of the waters. The night was spent in the open air, on the left bank.

On the 18th they set out at three in the morning, and near five in the afternoon reached the Raudal des Guahiboes, on the dike of which they landed while the Indians were drawing up the boat. The gneiss rock exhibited circular holes, produced by the friction of pebbles, in one of which they prepared a beverage consisting of water, sugar, and the juice of acid fruits, for the purpose of allaying the thirst of the missionary who was seized by a fever fit; after which they had the pleasure of bathing in a quiet place in the midst of the cataracts. After an hour's delay, the boat having been got up, they re-embarked their instruments and provisions. The river was 1705 yards broad, and had to be crossed obliquely, at a part where the waters rushed with extreme rapidity towards the bar over which they were precipitated. In the midst of this dangerous navigation they were overtaken by a thunderstorm accompanied by torrents of rain; and after rowing twenty minutes, found that so far from having made progress they were approaching the fall. But, as the Indians redoubled their efforts, the danger was escaped, and the boat arrived at nightfall in the port of Maypures. The night was extremely dark, and the village was at a considerable distance; still, as the missionary caused copal-torches to be lighted, they proceeded. As the rain ceased the zancudoes re-appeared, and the flambeaux being extinguished, they had to grope their way. One of their fellow-travellers, Don Nicolas Soto, slipped from a round trunk on which he attempted to cross a gully, but fortunately received no injury. To add to their distress, the pilot talked incessantly of venomous snakes, water-serpents, and tigers. On their arrival at the mission they found the inhabitants immersed in profound sleep, and nothing was heard but the cries of nocturnal birds and the distant roar of the cataract.

At the village of Maypures they remained three days, for the purpose of examining the neighbourhood. The cataract, called by the Indians Quittuna, is formed by an archipelago of islands, filling the bed of the river to the length of 6395 yards, and by dikes of rock which occasionally join them together. The largest of these shelves or bars are at Purimarimi, Manimi, and the Salto de la Sardina, the last of which is about nine feet high. To obtain a full view of the falls, the travellers frequently ascended the eminence of Manimi, a granitic ridge rising from the savannah, to the north of the church. " When one attains the summit of the rock," says Humboldt, " he suddenly sees a sheet of foam a mile in extent. Enormous masses of rock, of an iron blackness, emerge from its bosom, some of a mammillar form, and grouped like basaltic hills; others resembling towers, castles, and ruins. Their dark colour contrasts with the silvery whiteness of the foam. Every rock and islet is covered with tufts of stately trees. From the base of these prominences, as far as the eye can reach, there hangs over the river a dense mist, through which the tops of majestic palms are seen to penetrate. At every hour of the day this sheet of foam presents a different aspect. Sometimes the mountain isles and palms project their long shadows over it; sometimes the rays of the setting sun are refracted in the humid cloud that covers the cataract, when coloured arches form, vanish, and re-appear by turns."

The mountain of Manimi forms the eastern limit of a plain, which presented the same appearance as that of Atures. Toward the west is a level space formerly occupied by the waters of the river, and exhibiting rocks similar to the islands of the cataracts. These masses are also crowned with palms; and one of them, called Keri, is celebrated in the country for a white spot, which Humboldt supposed to be a large nodule of quartz. In an islet amidst the rush of waters there is a similar spot. The Indians view them with a mysterious interest, believing they see in the former the image of the moon, and in the latter that of the sun.

The inhabitants of the mission were Guahiboes and Macoes. In the time of the Jesuits the number was six hundred, but it had gradually fallen to less than sixty. They are represented as gentle, temperate, and cleanly. They cultivate plantains and cassava, and, like most of the Indians of the Orinoco, prepare nourishing drinks from the fruits of palms and other plants. Some of them were occupied in manufacturing a coarse pottery. Cattle, and especially goats, had at one time multiplied considerably at Maypures; but at the period of Humboldt's visit none were to be seen in any mission of the Orinoco. Tame macaws were seen round the huts, and flying in the fields like pigeons. Their plumage being of the most vivid tints of purple, blue, and yellow, these birds are a great ornament to the Indian farm-yards.

Round the village there grows a majestic tree of the genus *Unona*, with straight branches rising in the form of a pyramid. The infusion of the aromatic fruit is a powerful febrifuge, and is used as such in preference to the astringent bark of the *Cinchona* or *Bonplandia trifoliata*.

The longitude of this place was found to be 68° 17' 9", the latitude 5° 13' 57", differing from the best maps then existing by half a degree of longitude and as much of latitude. The thermometer during the night indicated from 80° to 84°, and in the day 86°. The water of the river was 81·7°, and that of a spring 82°.

Having spent some days at the mission of Maypures, the travellers embarked at two in the afternoon in the canoe procured at the turtle island, which, although considerably damaged by the carelessness of the Indians, was judged sufficient for the long voyage they had yet to perform. Above the great cataracts they found themselves as it were in a new world. Toward the east, in the extreme distance, rose the great chain of the Cunavami mountains, one of the peaks of which, named Calidamini, reflects at sunset a reddish glare of light. After encountering one more rapid they entered upon smooth water, and passed the night in a rocky island.

On the 22d they set out at an early hour. The morning was damp but delicious, and not a breath of wind was felt; a perpetual calm reigning to the south of the cataracts, which Humboldt attributes to the windings of the rivers, the shelter of mountains, and the almost incessant rains. In the valley of the Amazon, on the contrary, a strong breeze rises every day at two in the afternoon, which, however, is felt only along the line of the current. It always moves against the stream, and by means of it a boat may go up the Amazon under sail a length of 2590 miles. The great salubrity of this district is probably owing to the gale. They passed the mouths of several streams, and admired the grandeur of the cerros of Lipapo, a branch of the cordillera of Parime, the aspect of which varied every hour of the day. At sunrise, the dense vegetation with which they are covered was tinged with a darkgreen inclining to brown, while broad and deep shadows were projected over the neighbouring plain, forming a strong contrast with the vivid light diffused around. Toward noon the shadows disappeared, and the whole group was veiled in an azure vapour, which softened the outlines of the rocks, moderated the effects of light, and gave the landscape an aspect of calmness and repose. Landing at the mouth of the Rio Vichada to examine the vegetation, they found numberless small granitic rocks rising from the plain, and presenting the appearance of prisms, ruined columns, and towers. The forest was thin, and at the confluence of the two rivers, the rocks and even the soil were covered with mosses and lichens. M. Bonpland found several specimens of *Laurus cinnamomoides*, a very aromatic species of cinnamon, which, together with the American nutmeg, the pimento, and *Laurus pucheri*, Humboldt remarks, would have become important objects of trade, had not Europe, at the period when the New World was discovered, been already accustomed to the spices of India. The travellers rested at night on the bank of the Orinoco, at the mouth of the Zama. This river is one of those which are said to have black water, as it appears of a dark-brown or greenishblack; and here they entered the system of rivers to which the name of *Aguas Negras* is given. The colour is supposed to be owing to a solution of vegetable matter, and the Indians attribute it to the roots of sarsaparilla.

At five in the morning of the 23d they continued their voyage, and passed the mouth of the Rio Mataveni. The banks were still skirted by forests, but the mountains on the east retired farther back. The traces left by the floods were not higher than eight feet. At the place where they passed the night, multitudes of bats issued from the crevices, and hovered around their hammocks. Next day a violent rain obliged them to set out at a very early hour. In the afternoon they landed at the Indian plantations of San Fernando, and after midnight arrived at the mission, where they were received with the kindest hospitality.

The village of San Fernando de Atabipo is situated near the confluence of the Orinoco, the Atabipo, and the Guaviare; the latter of which Humboldt thinks might with more propriety be considered the continuation of the Orinoco than a branch. The number of inhabitants did not exceed 226. The missionary had the title of president of the stations on the Orinoco, and superintended the twenty-six ecclesiastics settled on its banks, as well as on those of the Rio Negro, Casiquiare, Atabipo, and Caura. The Indians were a little more civilized than the inmates of the other establishments, and cultivated cacao in small quantities, together with cassava and plantains. They were surrounded with good pasturage, but not more than seven or eight cows were to be seen. The most striking object in the neighbourhood was the pirijao palm, which has a thorny trunk more than sixty-four feet high, pinnated leaves, and clusters of fruits two or three inches in diameter, and of a purple colour. The fruit furnishes a farinaceous substance, of a colour resembling that of the yolk of an egg, which when boiled or roasted affords a very wholesome and agreeable aliment.

On entering the Rio Atabipo the travellers found

a great change in the scenery, the colour of the stream, and the constitution of the atmosphere. The trees were of a different species; the mosquitoes had entirely disappeared, and the waters, instead of being turbid, and loaded with earthy matter, were of a dark colour, clear, agreeable to the taste, and two degrees cooler. So great is their transparency, that the smallest fishes are distinguishable at the depth of twenty or thirty feet, and the bottom, which consists of white quartzy sand, is usually visible. The banks covered with plants, among which rise numerous palms, are reflected by the surface of the river with a vividness almost as bright as that of the objects themselves. Above the mission no crocodiles occur, but their place is supplied by bavas and fresh-water dolphins. The chiguires, howling-monkeys, and zamuro-vultures had disappeared, though jaguars were still seen, and the water-snakes were extremely numerous.

On the 26th the travellers advanced only two or three leagues, and passed the night on a rock near the Indian plantations of Guapasoso. At two in the morning they again set out, and continued to ascend the river. About noon they passed the granitic rock named Piedra del Tigre, and at the close of the day had great difficulty in finding a suitable place for sleeping, owing to the inundation of the banks. It rained hard from sunset, and as the missionary had a fit of tertian fever they re-embarked immediately after midnight. At dawn they landed to examine a gigantic ceiba-tree, which was nearly 128 feet in height, with a diameter of fifteen or sixteen feet. On the 29th the air was cooler, but loaded with vapours, and the current being strong they advanced slowly. It was night when they arrived at the mission of San Baltasar, where they lodged with a Catalan priest, a lively and agreeable person. The village was built with great regularity, and the plantations seemed better cultivated than elsewhere.

At a late hour in the morning they left his abode, and after ascending the Atabipo for five miles entered the Rio Temi. A granitic rock on the western bank of the former river attracted their attention. It is called the Piedra de la Guahiba or Piedra de la Madre, and commemorates one of those acts of oppression of which Europeans are guilty in all countries whenever they come into contact with savages. In 1797, the missionary of San Fernando had led his people to the banks of the Rio Guaviare on a hostile excursion. In an Indian hut they found a Guahibo woman, with three children, occupied in preparing cassava-flour. She and her little ones attempted to escape, but were seized and carried away. The unhappy female repeatedly fled with her children from the village, but was always traced by her Christian countrymen. At length the friar, after causing her to be severely beaten, resolved to separate her from her family, and sent her up the Atabipo toward the missions of the Rio Negro. Ignorant of the fate intended for her, but judging by the direction of the sun that her persecutors were carrying her far from her native country, she burst her fetters, leaped from the boat, and swam to the left bank of the river. She landed on a rock; but the president of the establishment ordered the Indians to row to the shore and lay lands on her. She was brought back in the evening, stretched upon the bare stone (the Piedra de la Madre), scourged with straps of manatee leather, which are the ordinary whips of the country, and then dragged to the mission of Javita, her hands bound behind her back. It was the rainy season, the night was excessively dark, forests believed to be impenetrable stretched from that station to San Fernando over an extent of 86 miles, and the only communication between these places was by the river; yet the Guahibo mother, breaking her bonds, and eluding the vigilance of her guards, escaped under night, and on the fourth morning was seen at the village, hovering around the hut which contained her children. On this journey she must have undergone hardships from which the most robust man would have shrunk; was forced to live upon ants, to swim numerous streams, and to make her way through thickets and thorny lianas. And the reward of all this courage and devotion was—her removal to one of the missions of the Upper Orinoco, where, despairing of ever seeing her beloved children, and refusing all kind of nourishment, she died, a victim to the bigotry and barbarity of wretches blasphemously calling themselves the ministers of a religion which inculcates universal benevolence.

Above the mouth of the Guasucavi the travellers entered the Rio Temi, which runs from south to north. The ground was flat and covered with trees, over which rose the pirijao palm with its clusters of peach-like fruits, and the *Mauritia aculeata,* with fan-shaped leaves pointing downwards, and marked with concentric circles of blue and green. Wherever the river forms sinuosities the forest is flooded to a great extent; and, to shorten the route, the boat frequently pushed through the woods along open avenues of water four or five feet broad. An Indian furnished with a large knife stood at the bow continually cutting the branches which obstructed the passage. In the thickest part of it a shoal of freshwater dolphins issued from beneath the trees and surrounded the vessel. At five in the evening the travellers, after sticking for some time between two trunks and experiencing other difficulties, regained the proper channel, and passed the night near one of the columnar masses of granite which occasionally protrude from the level surface.

Setting out before daybreak, they remained in the bed of the river till sunrise, when, to avoid the force of the current, they again entered the inundated forest; and soon arriving at the junction of the Temi with the Tuamini, they followed the latter toward the south-west. At eleven they reached San Antonio de Javita, where they had the pleasure of finding a very intelligent and agreeable monk: though they were obliged to remain nearly a week while the boat was carried by land to the Rio Negro. For two days the travellers had felt an extraordinary irritation on the joints of the fingers and on the back of the hands, which the missionary informed them was caused by insects. Nothing could be distinguished with a lens but parallel streaks of a whitish colour, the form of which has obtained for these animalculæ the name of *aradores*, or ploughmen. A mulatto woman engaged to extirpate them one by one, and, digging with a small bit of pointed wood, at length succeeded in extracting a little round bag; but Humboldt did not possess sufficient patience to wait for relief from so tedious an operation. Next day, however, an Indian effected a radical cure by means of the infusion of bark stripped from a certain shrub.

In 1755, before the expedition to the boundaries, the country between the missions of Javita and San Baltasar was dependent on Brazil, and the Portuguese had advanced from the Rio Negro as far as the banks of the Temi. An Indian chief named Javita, one of their auxiliaries, pushed his hostile excursions to a distance of more than 345 miles; and, being furnished with a patent for drawing the natives from the forest " for the conquest of souls," did not fail to make use of it for selling slaves to his allies. When Solano, one of the leaders of the expedition just described, arrived at San Fernando de Atabipo, he seized the adventurer, and by treating him with gentleness gained him over to the interests of the Spaniards. He was still living when the travellers proceeded to the Rio Negro; and, as he attended them on all their botanical excursions, they obtained much information from him. He assured them, that he had seen almost all the Indian tribes which inhabit the vast countries between the Upper Orinoco, the Rio Negro, the Irinida, and the Jupura, devour human flesh. Their cannibalism he considered as the effect of a system of revenge, as they eat only enemies who are made prisoners in battle.

The climate of the mission of San Antonio de Javita is so rainy that the sun and stars are seldom to be seen, and the padre informed the travellers that it sometimes rained without intermission for four or five months. The water that fell in five hours on the 1st of May, Humboldt found to be 21 lines in height, and on the 3d of May he collected 14 lines in three hours; whereas at Paris there fall only 28 or 30 lines in as many weeks. The temperature is lower than at Maypures, but higher than on the Rio Negro; the thermometer standing at 80° or 80·6° by day, and at 69·8° by night.

The Indians of the mission amounted only to 160. Some of them were employed in the construction of boats, which are formed of the trunks of a species of laurel (*Ocotea cymbarum*), hollowed by means of fire and the axe. These trees attain a height of more than a hundred feet, and have a yellow resinous wood which emits an agreeable odour. The forest between Javita and Pimichin affords an immense quantity of gigantic timber, as tall occasionally as 116 or 117 feet; but as the trees give out branches only towards the summit, the travellers were disappointed, amid so great a profusion of unknown species, in not being able to procure the leaves and flowers. Besides, as it rained incessantly so long a time, M. Bonpland lost the greater part of his dried specimens. Although no pines or firs occur in these woods, balsams, resins, and aromatic gums, are abundantly furnished by many other trees, and are collected as objects of trade by the people of Javita.

At the mission of San Baltasar they had seen the natives preparing a kind of elastic gum, which they said was found under ground; and in the forests at Javita, the old Indian who accompanied them showed that it was obtained by digging several feet deep among the roots of two particular trees, the *Hevea* of Aublet and one with pinnate leaves. This substance, which bears the name of dapicho, is white, corky, and brittle, with a laminated structure and undulating edges; but on being roasted it assumes a black colour, and acquires the properties of caoutchouc.

The natives of these countries live in hordes of forty or fifty, and unite under a common chief only when they wage war with their neighbours. As the different tribes speak different languages they have little communication. They cultivate cassava, plantains, and sometimes maize; but shift from place to place, so that they entirely lose the advantages resulting in other countries from agricultural habits. They have two great objects of worship,—the good principle, Cachimana, who regulates the seasons and favours the harvests; and the evil principle, Jolokiamo, less powerful, but more active and artful. They have no idols; but the botuto, or sacred trumpet, is an object of veneration, the initiation into the mysteries of which requires pure manners and a single life. Women are not permitted to see it, and are excluded from all the ceremonies of this religion.

It took the Indians more than four days to drag the boat upon rollers to the Rio Pimichin. One of them, a tall strong man, was bitten by a snake, and was brought to the mission in a very alarming condition. He had dropped down senseless, and was afterwards seized with nausea, vertigo, and a determination of blood to the head, but was cured by an infusion of raiz de mato; respecting the plant furnishing which Humboldt could obtain no satisfactory information, although he supposes it to be of the family of Apocyneæ. In the hut of this individual he observed balls of an earthy and impure salt, two or three inches in diameter. It is obtained by reducing to ashes the spadix and fruit of a palm-tree, and consists of muriate of potash and soda, caustic lime, and other ingredients. The Indians dissolve a few grains in water, which they drop on their food.

On the 5th May the travellers set off on foot to follow their canoe. They had to ford numerous streams, the passage of which was somewhat dangerous on account of the number of snakes in the marshes. After passing through dense forests of lofty trees, among which they noted several new species of coffee and other plants, they arrived toward evening at a small farm on the Pimichin, where they passed the night in a deserted hut, not without apprehension of being bitten by serpents, as they were obliged to lie on the floor. Before they took possession of this shed their attendants killed two great Mapanare snakes, and in the morning a large viper was found beneath the jaguar-skin on which one of them had slept. This species of serpent is white on the belly, spotted with brown and black on the back, and grows to the length of four or five feet. Humboldt remarks, that if vipers and rattlesnakes had such a disposition for offence as is usually supposed, the human race could not have resisted them in some parts of America.

Embarking at sunrise they proceeded down the Pimichin, which is celebrated for the number of its windings. It is navigable during the whole year, and has only one rapid. In four hours and a half they entered the Rio Negro. "The morning," says Humboldt, " was cool and beautiful; we had been confined thirty-six days in a narrow canoe, so unsteady that it would have been overset by any one rising imprudently from his seat, without warning the rowers to preserve its balance by leaning to the opposite side. We had suffered severely from the stings of insects, but we had withstood the insalubrity of the climate; we had passed without accident the numerous falls and bars that impede the navigation of the rivers, and often render it more dangerous than long voyages by sea.

"After all that we had endured, I may be allowed to mention the satisfaction which we felt in having reached the tributaries of the Amazon,—in having passed the isthmus which separates two great systems of rivers,—and in having attained a certainty of fulfilling the most important object of our journey,—that of determining by astronomical observations the course of that arm of the Orinoco which joins the Rio Negro, and whose existence had been alternately proved and denied for half a century. In these inland regions of the New Continent we almost accustom ourselves to consider man as inessential to the order of nature. The earth is overloaded with plants, of which nothing impedes the development. An immense layer of mould evinces the uninterrupted action of the organic powers. The crocodiles and boas are masters of the river; the jaguar, pecari, dante, and monkeys of numerous species, traverse the forest without fear and without danger, residing there as in an ancient heritage. On the ocean and on the sands of Africa, we with difficulty reconcile ourselves to the disappearance of man; but here his absence, in a fertile country clothed with perpetual verdure, produces a strange and melancholy feeling."

The Rio Negro, which flows eastward into the Amazon, was for ages considered of great political importance by the Spanish government, as it would have furnished to the Portuguese an easy introduction into the missions of Guiana. The jealousies of these rival nations, the ignorance and diversified languages of the Indians, the difficulty of penetrating into these inland regions, and other causes, rendered the knowledge of the sources as well as the tributaries of the Negro and Orinoco extremely defective. To endeavour to throw some light on this geographical point, and in particular to determine the course of that branch of the Orinoco which joins the Rio Negro, was the great object of Humboldt's journey. This last, or Black River, is so named on account of the dark colour of its waters, which are of an amber hue wherever it is shallow, and dark-brown wherever the depth is great. After entering it by the Pimichin, and passing the rapid at the confluence of the two streams, the travellers soon reached the mission of Maroa, containing 150 Indians, where they purchased some fine toucans. Passing the station of Tomo they visited that of Davipe, where they were received by the missionary with great hospitality. Here they bought some fowls and a pig, which interested their servants so much that they pressed them to depart, in order to reach the island of Dapa where the animal might be roasted. They arrived at sunset, and found some cultivated ground and an Indian hut. Four natives were seated round a fire eating a kind of paste, consisting of large ants, of which several bags were suspended over the fire. There were more than fourteen persons in this small cabin, lying naked in hammocks placed above each other. They received Father Zea with great joy, and two young women prepared cassava-cakes; after which the travellers retired to rest. The family slept only till two in the morning, when they began to converse in their hammocks. This custom of being awake four or five hours before sunrise Humboldt found to be general among the people of Guiana; and, hence, when an attempt is made to surprise them, the first part of the night is chosen for the purpose.

Proceeding down the Rio Negro they passed the mouth of the Casiquiare, the river by which a communication is effected between the former and the Orinoco; and towards evening reached the mission of San Carlos del Rio Negro, with the commander of which they lodged. The military establishment of this frontier post consisted of seventeen soldiers, ten of whom were detached for the security of the neighbouring stations. The voyage from the mouth of the Rio Negro to Grand Para occupying only twenty or twenty-five days, it would not have taken much more time to have gone down the Amazon to the coast of Brazil, than to return by the Casiquiare and Orinoco to that of Caraccas; but our travellers were informed that it was difficult to pass from the Spanish to the Portuguese settlements; and it was well for them that they declined this route, for they afterwards learned that instructions had been issued to seize and convey them to Lisbon. This project, however, was not countenanced by the government at home, who, when informed of the zeal of its subaltern agents, gave instant orders that the philosophers should not be disturbed in their pursuits.

Among the Indians of the Rio Negro they found some of those green pebbles known by the name of Amazon-stones, and which are worn as amulets.

The form usually given to them is that of the Persepolitan cylinders longitudinally perforated. These hard substances denote a degree of civilisation superior to that of the present inhabitants, who, so far from being able to cut them, imagine that they are naturally soft when taken out of the earth, and harden after they have been moulded by the hand. They were found to be jade or saussurite, approaching to compact felspar, of a colour passing from apple to emerald green, translucent on the edges, and taking a fine polish; but the substance usually called Amazon-stone in Europe is different, being a common felspar of a similar colour, coming from the Uralian Mountains and Lake Onega in Russia.

Connected with this mineral are the warlike women, whom the travellers of the sixteenth century named the Amazons of the New World; and regarding whom Humboldt found no satisfactory accounts, although he is disposed to believe that their existence was not merely imaginary.

The travellers passed three days at San Carlos, watching the greater part of each night, in the hope of seizing the moment of the passage of some star over the meridian; but the sky was continually obscured by vapours. On the 10th May they embarked a little before sunrise to go up the Rio Negro. The morning was fine, but as the heat increased the firmament became darkened. Passing between the islands of Zaruma and Mibita, covered with dense vegetation, and ascending the rapids of the Piedra de Uinumane, they entered the Casiquiare at the distance of $9\frac{1}{2}$ miles from the fort of San Carlos. The rock at the rapids was granite, traversed by numerous veins of quartz several inches broad. The night was spent at the mission of San Francisco Solano, on the left bank of the Casiquiare. The Indians were of two nations, the Pacimonales and Cheruvichahenas; and from the latter the travellers endeavoured to obtain some information respecting the upper part and sources of the Rio Negro, but without success. In one of the huts of the former tribe they purchased two large birds, a toucan and a macaw, to add to the already considerable stock which they possessed. Most of the animals were confined in small cages, while others ran at liberty all over the boat. At the approach of rain, the macaws uttered frightful screams, the toucan was desirous of gaining the shore in order to fish, and the little monkeys went in search of Father Zea to obtain shelter in his large sleeves. At night the leather case containing their provisions was placed in the centre; then the instruments and cages; around which were suspended the hammocks of the travellers; and beyond them the Indians slept, protected by a circle of fires to keep off the jaguars.

On the 11th they left the mission of San Francisco Solano at a late hour to make a short day's journey, for the vapours had begun to break up, and the travellers were unwilling to go far from the mouth of the Casiquiare without determining the longitude and latitude. This they had an opportunity of doing at night in the neighbourhood of a solitary granite rock, the Piedra di Culimacari, which they found to be in lat. 2° 0′ 42″ north, and long. 67° 13′ 26″ west. The determination was of great importance in a geographical and political point of view, for the greatest errors existed in maps, and the equator had been considered as the boundary between the Spanish and Portuguese possessions.

Leaving the Rock of Culimacari at half after one in the morning, they proceeded against the current, which was very rapid. The waters of the Casiquiare are white, and the mosquitoes again commenced their invasions, becoming more numerous as the boat receded from the black stream of the Rio Negro. In the whole course of the Casiquiare they did not find in the Christian settlements a population of 200 individuals, and the free Indians have retired from its banks. During a great part of the year the natives subsist on ants. At the mission of Mandavaca, which they reached in the evening, they found a monk who had spent twenty years in the country, and whose legs were so spotted by the stings of insects that the whiteness of the skin could scarcely be perceived. He complained of his solitude, and the sad necessity which often compelled him to leave the most atrocious crimes unpunished. An indigenous alcayde, or overseer, had a few years before eaten one of his wives, after fattening her by good feeding. " You cannot imagine," said the missionary, "all the perversity of this Indian family. You receive men of a new tribe into the village; they appear to be good, mild, and industrious; but suffer them to take part in an incursion to bring in the natives, and you can scarcely prevent them from murdering all they meet, and hiding some portions of the dead bodies." The travellers had in their canoe a fugitive Indian from the Guaisia, who in a few weeks had become sufficiently civilized to be very useful. As he was mild and intelligent, they had some desire of taking him into their service;

but discovering that his anthropophagous propensities remained they gave up the idea. He told them that "his relations (the people of his tribe) preferred the inside of the hands in man, as in bears," accompanying the assertion with gestures of savage joy.

Although the Indians of the Casiquiare readily return to their barbarous habits, they manifest, while in the missions, intelligence, industry, and a great facility in learning the Spanish tongue. As the villages are usually inhabited by three or four tribes who do not understand each other, the language of their instructor affords a general means of communication. The soil on the Casiquiare is of excellent quality. Rice, beans, cotton, sugar, and indigo, thrive wherever they have been tried; but the humidity of the air, and the swarms of insects, oppose almost insuperable obstacles to cultivation. Immense bands of white ants destroy every thing that comes in their way, insomuch, that when a missionary would cultivate salad or any European culinary vegetable, he fills an old boat with soil, and having sown the seeds suspends it with cords, or elevates it on posts.

From the 14th to the 21st the travellers continued to ascend the Casiquiare, which flowed with considerable rapidity, having a breadth of 426 yards, and bordered by two enormous walls of trees hung with lianas. No openings could be discovered in these fences; and at night the Indians had to cut a small spot with their hatchets to make room enough for their beds, it being impossible to remain in the canoe on account of the mosquitoes and heavy rains. Great difficulty was experienced in finding wood to make a fire, the branches being so full of sap that they would scarcely burn. On shore the pothoses, arums, and lianas, furnished so thick a covering, that although it rained violently they were completely sheltered. At their last resting-place on the Casiquiare, the jaguars carried off their great dog while they slept.

On the 21st May they again entered the channel of the Orinoco, three leagues below the mission of Esmeralda. Here the scenery wore a very imposing aspect, lofty granitic mountains rising on the northern bank. The celebrated bifurcation of the river takes place in this manner: The stream, issuing from among the mountains, reaches the opening of a valley or depression of the ground which terminates at the Rio Negro, and divides into two branches. The principal branch continues its course toward the west-north-west, turning round the group of the mountains of Parime, while the other flows off southward and joins the Rio Negro. By this latter branch our travellers ascended from the river just mentioned, and again entered the Orinoco, four weeks after they had left it near the mouth of the Guaviare. They had still a voyage of 863 miles to perform before reaching Angostura.

CHAPTER XIX.

Route from Esmeralda to Angostura.

Mission of Esmeralda—Curare Poison—Indians—Duida Mountain—Descent of the Orinoco—Cave of Ataruipe—Raudalito of Carucari—Mission of Uruana—Character of the Otomacs—Clay eaten by the Natives—Arrival at Angostura—The Travellers attacked by Fever—Ferocity of the Crocodiles.

Opposite the point where the division of the river takes place, there rises in the form of an amphitheatre a group of granitic mountains, of which the principal one bears the name of Duida. It is about 8500 feet high; and being perpendicular on the south and west, bare and stony on the summit, and clothed on its less steep declivities with vast forests, presents a magnificent spectacle. At the foot of this huge mass is placed the most solitary and remote Christian settlement on the Upper Orinoco,—the mission of Esmeralda, containing eighty inhabitants. It is surrounded by a beautiful plain, covered with grasses of various species, pine-apples, and clumps of Mauritia palm, and watered by limpid rills.

There was no monk at the village; but the travellers were received with kindness by an old officer, who, taking them for Catalonian shopkeepers, admired their simplicity when he saw the bundles of paper in which their plants were preserved, and which he supposed they intended for sale. Notwithstanding the smallness of the mission three Indian languages were spoken in it; and among the inhabitants were some Zamboes, mulattoes, and copper-coloured people. A mineralogical error gave celebrity to Esmeralda, the rock-crystals and chloritic quartzes of Duida having been mistaken for diamonds and emeralds. The converts live in great poverty, and their misery is augmented by prodigious swarms of mosquitoes. Yet the situation of the establishment is exceedingly picturesque; the surrounding country is possessed of great fertility; and plantains, indigo, sugar, and cacao, might be produced in abundance.

This village is the most celebrated spot on the Orinoco for the manufacture of the curare, a very active poison employed in war and in the chase, as well as a remedy for gastric obstructions. Erroneous ideas had been entertained of this substance; but our travellers had an opportunity of seeing it prepared. When they arrived at Esmeralda, most of the Indians had just finished an excursion to gather juvias or the fruit of the bertholletia, and the liana which yields the curare. Their return was celebrated by a festival, which lasted several days, during which they were in a state of intoxication. One less drunk than the rest was employed in preparing the poison. He was the chemist of the place, and boasted of his skill, extolling the composition as superior to any thing that could be made in Europe. The liana which yields it is named bejuco, and appeared to be of the Strychnos family. The branches are scraped with a knife, and the bark that comes off is bruised, and reduced to very

R

thin filaments on the stone employed for grinding cassava. A cold infusion is prepared by pouring water on this fibrous mass, in a funnel made of a plantain-leaf rolled up in the form of a cone, and placed in another somewhat stronger made of palm-leaves, the whole supported by a slight framework. A yellowish fluid filters through the apparatus. It is the venomous liquor; which, however, acquires strength only when concentrated by evaporation in a large earthen pot. To give it consistence, it is mixed with a glutinous vegetable juice, obtained from a tree named kiracaguera. At the moment when this addition is made to the fluid, now kept in a state of ebullition, the whole blackens, and coagulates into a substance resembling tar or thick syrup. The curare may be tasted without danger; for, like the venom of serpents, it only acts when introduced directly into the blood, and the Indians consider it as an excellent stomachic. It is universally employed by them in hunting, the tips of their arrows being covered with it; and the usual mode of killing domestic fowls is to scratch the skin with one of these infected weapons. Other species of vegetable poison are manufactured in various parts of Guiana.

After seeing this composition prepared, the philosophers accompanied the artist to the festival of the juvias. In the hut where the revellers were assembled, large roasted monkeys blackened by smoke were ranged against the wall. Humboldt imagines that the habit of eating animals so much resembling man has in some degree contributed to diminish the horror of anthropophagy among savages. Apes when thus cooked, and especially such as have a very round head, bear a hideous likeness to a child; and for this reason such Europeans as are obliged to feed upon them separate the head and hands before the dish is presented at their tables. The flesh is very lean and dry.

Among the articles brought by the Indians from their expedition were various interesting vegetable productions; fruits of different species, reeds upwards of fifteen feet long, perfectly straight and free of knots, and bark used for making shirts. The women were employed in serving the men with the food already mentioned, fermented liquors, and palm-cabbage, but were not permitted to join in the festivities. Among all the tribes of the Orinoco the females live in a sort of slavery, almost the whole labour devolving upon them. Polygamy is frequently practised, and on the other hand a kind of polyandry is established in places where the fair sex are less numerous. When a native who has several wives becomes a Christian, the missionaries compel him to choose her whom he prefers and to dismiss the others.

The summit of Duida is so steep that no person has ever ascended it. At the beginning and end of the rainy season, small flames, which appear to shift, are seen upon it. On this account the mountain has been called a volcano, which, however, it is not. The granite whereof it is composed is full of veins, some of which being partly open, gaseous and inflammable vapours may pass through them; for it is not probable that the flames are caused by lightning, the humidity of the climate being such that plants do not readily take fire.

The travellers had an opportunity of seeing at Esmeralda some of the dwarf and fair Indians, that ancient traditions had mentioned as living near the sources of the Orinoco. The Guaicas, or diminutive class, whom they measured, were in general from 4 feet 10½ to 4 feet 11½ inches in height; and it was said that the whole tribe was of the same stature. The Guahariboes, or fair variety, were similar to the others in form and features, and differed only in having the skin of a lighter tint.

On the 23d May the travellers left the mission of Esmeralda in a state of languor and weakness, caused by the torment of insects, bad nourishment, and a long voyage performed in a narrow and damp boat. They had not attempted to ascend the Orinoco towards its sources, as the country above that station was inhabited by hostile Indians; so that of the two geographical problems connected with the river,—the position of its sources and the nature of its communication with the Rio Negro,—they had been obliged to content themselves with the solution of the latter. When they embarked they were surrounded by the mulattoes and others who considered themselves Spaniards, and who entreated them to solicit from the governor of Angostura their return to the Llanos, or at least their removal to the missions of the Rio Negro. Humboldt pleaded the cause of these proscribed men at a subsequent period; but his efforts were fruitless. The weather was very stormy, and the summit of Duida was enveloped in clouds; but the thunders which rolled there did not disturb the plains. Nor did they, generally speaking, observe in the valley of the Orinoco those violent electric explosions which almost every night, during the rainy season, alarm the traveller along the Rio Magdalena. After four hours' navigation in descending the stream, they arrived at the bifurcation, and reposed on the same beach of the Casiquiare where, a few days before, their dog had been carried off by the jaguars. The cries of these animals were again heard through the whole night. The black tiger also occurs in these districts. It is celebrated for its strength and ferocity, and appears to be larger than the other, of which, however, it is probably a variety.

Leaving their resting-place before sunrise, and sailing with the current, they passed the mouths of the Cunucunumo, Guanami, and Puruname. The country was entirely desert, although rude figures representing the sun, the moon, and different animals, are to be seen on the granite rocks; attesting the former existence of a people more civilized than any that they had seen.

On the 27th May they reached the mission of San Fernando de Atabipo, where they had lodged a month before on their ascent toward the Rio Negro. The president had allowed himself to become very uneasy respecting the object of their journey; and requested Humboldt to leave a writing in his hands, bearing testimony to the good order that prevailed in the Christian settlements on the Orinoco, and the mildness with which the natives were treated. This, however, he declined. From this point they retraced their former route, and passed the cataracts. On the 31st, they landed before sunset at the Puerto de la Expedicion, for the purpose of visiting the cave of Ataruipe, which is the sepulchre of an extinct nation.

"We climbed," says Humboldt, "with difficulty and not without danger, a steep rock of granite, entirely destitute of soil. It would have been almost impossible to fix the foot on this smooth and highly-inclined surface, had not large crystals of felspar, which had resisted decomposition, projected from the rock so as to present points of support. Scarcely had we reached the summit of the mountain when we were struck with astonishment at the extraordinary appearance of the surrounding country: The foamy bed of the waters was filled with an archipelago of islands covered with palms. Toward the west, on the left bank of the Orinoco extended the savannahs of the Meta and Casanare, like a sea of verdure, the misty horizon of which was illuminated by the rays of the setting sun. The mighty orb, like a globe of fire suspended over the plain, and the solitary peak of Uniana, which appeared more lofty from being wrapped in vapours that softened its outlines, contributed to impress a character of sublimity upon the scene. We looked down into a deep valley enclosed on every side. Birds of prey and goatsuckers winged their solitary way in this inaccessible circus. We found pleasure in following their fleeting shadows as they glided slowly over the flanks of the rock.

" A narrow ridge led us towards a neighbouring mountain, the rounded summit of which supported enormous blocks of granite. These masses are more than 40 or 50 feet in diameter, and present a form so perfectly spherical, that, as they seem to touch the ground only by a small number of points, it might be supposed that the slightest shock of an earthquake would roll them into the abyss. I do not remember to have seen anywhere else a similar phenomenon amid the decompositions of granitic deposites. If the balls rested upon a rock of a different nature, as is the case with the blocks of Jura, it might be supposed that they had been rounded by the action of water, or projected by the force of an elastic fluid; but their position on the summit of a hill of the same nature, renders it more probable that they owe their origin to a gradual decomposition of the rock.

" The most remote part of the valley is covered by a dense forest. In this shady and solitary place, on the declivity of a steep mountain, opens the cave of Ataruipe. It is less a cave than a projecting rock, in which the waters have scooped a great hollow, when, in the ancient revolutions of our planet, they had reached to that height. In this tomb of a whole extinct tribe we soon counted nearly 600 skeletons in good preservation, and arranged so regularly that it would have been difficult to make an error in numbering them. Each skeleton rests upon a kind of basket formed of the petioles of palms. These baskets, which the natives call *mapires*, have the form of a square bag. Their size is proportional to the age of the dead; and there are even some for infants which had died at the moment of birth. We saw them from ten inches and a half to three feet six inches and a half in length. All the skeletons are bent, and so entire that not a rib or a bone of the fingers or toes is wanting. The bones have been prepared in three different ways,—whitened in the air and sun, dyed red with onoto, a colouring matter obtained from the *Bixa orellana;* or, like mummies, covered with odorous resins, and enveloped in leaves of heliconia and banana. The Indians related to us that the corpse is first placed in the humid earth, that the flesh may be consumed by degrees. Some months after, it is taken out, and the flesh that remains on the bones is scraped off with sharp stones. Several tribes of Guiana still follow this practice. Near the mapires or baskets there were vases of half-burnt clay, which appeared to contain the bones of the same family. The largest of these vases or funereal urns are three feet two inches high, and four feet six inches long. They are of a greenish-gray colour, and have an oval form, not unpleasant to the eye. The handles are made in the form of crocodiles or serpents, and the edge is encircled by meanders, labyrinths, and grecques, with narrow lines variously combined. These paintings are seen in all countries, among nations placed at the greatest distances from each other, and the most different in respect to civilisation. The inhabitants of the little mission of Maypures execute them at the present day on their most common pottery. They adorn the shields of the Otaheitans, the fishing-instruments of the Esquimaux, the walls of the Mexican palace of Mitla, and the vases of Magna Græcia.

" We opened, to the great concern of our guides, several mapires, for the purpose of attentively examining the form of the skulls. They all presented the characters of the American race,—two or three only approached the Caucasian form. We took several skulls, the skeleton of a child of six or seven years, and those of two full-grown men, of the nation of the Atures. All these bones, some painted red, others covered with odorous resins, were placed in the mapires or baskets already described. They formed nearly the whole lading of a mule; and, as we were aware of the superstitious aversion which the natives show towards dead bodies, after they have given them burial, we carefully covered the baskets with new mats. Unfortunately for us, the penetration of the Indians, and the extreme delicacy of their organs of smell, rendered our precautions useless. Wherever we stopped,—in the Carib missions, in the midst of the Llanos, between Angostura and New Barcelona,—the natives collected around our mules to admire the monkeys which we had brought from the Orinoco. These good people had scarcely touched our baggage when they predicted the approaching death of the beast of burden 'that carried the dead.' In vain we told them that they were deceived in their conjectures, that the panniers contained bones of crocodiles and lamantins; they persisted in repeating, that they smelt the resin which surrounded the skeletons, and that 'they were some of their old relatives.'

" We departed in silence from the cave of Ataruipe. It was one of those calm and serene nights which are so common in the torrid zone. The stars shone with a mild and planetary light; their scintillation was scarcely perceptible at the horizon, which seemed illuminated by the great nebulæ of the southern hemisphere. Multitudes of insects diffused a reddish light over the air. The ground, profusely covered with plants, shone with those living and moving lights as if the stars of the firmament had fallen upon the savannah. On leaving the cave, we repeatedly stopped to admire the beauty of this extraordinary place. The scented vanilla and festoons of bignoniæ decorated its entrance; while the summit of the overhanging hill was crowned by arrowy palm-trees that waved murmuring in the air."

Similar caves are said to exist to the north of the cataracts; but the tombs of the Indians of the Orinoco have not been sufficiently examined, because they do not, like those of Peru, contain treasures.

The travellers staid at the mission of Atures only so long as was necessary for the passage of their canoe through the great falls. The priest, Bernardo Zea, who had accompanied them to the Rio Negro, remained behind. His ague had not been removed; but its attacks had become an habitual evil, to which he now paid little attention. Fevers of a more destructive kind prevailed in the establishment, insomuch that the greater part of the inmates were confined to their hammocks. Again embarked on the Orinoco the travellers ventured to descend the lower half of the rapids of Atures, landing here and there to climb the rocks, among which the golden manakin (*Pipra rupicola*), one of the most beautiful birds of the tropics, builds its nest. At the Raudalito of Carucari, they entered some of the caverns formed by the piling up of granite blocks, and enjoyed the extraordinary spectacle of the river dashing in a sheet of foam over their heads. The boat was to coast the eastern bank of a narrow island, and take them in after a long circuit; but it did not make its appearance, and night approaching, together with a tremendous thunder-storm, M. Bonpland was desirous of swimming across, in order to seek assistance at Atures from Father Zea. Humboldt and the other person who was with them dissuaded him with difficulty from so hazardous an enterprise; and shortly after two large crocodiles made their appearance, attracted by the plaintive cries of the monkeys. At length the Indians arrived with the vessel, and the navigation was continued during part of the night. At Carichana the missionary received them with kindness. Here the travellers remained some days to recruit their exhausted strength, and M. Bonpland had the satisfaction of dissecting a manatee.

From Carichana they went in two days to the mission of Uruana, the situation of which is extremely picturesque, the village being placed at the foot of a lofty granitic mountain, the columnar rocks appearing at intervals above the trees. Here the river is more than 4263 yards broad, and runs in a straight line directly east. The hamlet is inhabited by the Otomacs, one of the rudest of the American tribes. These Indians swallow quantities of earth for the purpose of allaying hunger. When the waters are low they live on fish and turtles; but when the rivers swell, and it becomes difficult to procure that food, they eat daily a large portion of clay. The travellers found in their huts heaps of it in the form of balls, piled up in pyramids three or four feet high. This substance is fine and unctuous, of a yellowish-gray colour, containing silica and alumina, with three or four per cent. of lime. Being a restless and turbulent people, with unbridled passions and excessively given to intoxication, the little village of Uruana is more difficult to govern than any of the other missions. By inhaling at the nose the powder obtained from the pods of the *Acacia niopo* they throw themselves into a state of intoxication bordering on madness, that lasts several days, during which dreadful murders are committed. The most vindictive cover the nail of the thumb with the curare poison, the slightest scratch being thus sufficient to produce death. When this crime is perpetrated at night they throw the body into the river. "Every time," said the monk, "that I see the women fetch water from a part of the shore to which they do not usually go for it, I suspect that a murder has been committed in my mission."

On the 7th June the travellers took leave of Father Ramon Bueno, whom Humboldt eulogizes as the only one of ten missionaries of Guiana whom they had seen who appeared to be attentive to any thing that regarded the natives. The night was passed at the island of Cucurupara, to the east of which is the mouth of the Cano de la Tortuga. On its southern bank is the almost deserted station of San Miguel de la Tortuga, in the neighbourhood of which, according to the Indians, are otters with a very fine fur, and lizards with two feet.

From the island of Cucurupara to Angostura the capital of Guiana, a distance of little less than 328 miles, the travellers were only nine days on the water. On the 8th June they landed at a farm opposite the mouth of the Apure, where Humboldt obtained some good observations of latitude and longitude; and on the 9th met a great number of boats laden with goods, on their way to that river. Here Don Nicolas Soto, who had accompanied them on their voyage to the Rio Negro, took leave and returned to his family. As they advanced the population became more considerable, consisting almost exclusively of whites, negroes, and mulattoes. On the 11th they passed the mouth of the Rio Caura, near which is a small lake formed in 1790 by the sinking of the ground in consequence of an earthquake. The Boca del Infierno and the Raudal de Camiseta, a series of whirlpools and rapids caused by a chain of small rocks, were the only remarkable features that occurred until they reached Angostura.

On arriving at the capital, they hastened to present themselves to Don Felipe de Ynciarte the governor of Guiana, who received them in the most obliging manner. A painful circumstance forced them to remain a whole month in this place. They were both, a few days after their arrival, attacked by a disorder, which in M. Bonpland assumed the character of a typhoid fever. A mulatto servant, who had attended them from Cumana, was similarly affected. His death was announced on the ninth day; but he had only fallen into a state of insensibility which lasted several hours, and was followed by a salutary crisis. Humboldt escaped with a very violent attack, during which he was made to take a mixture of honey and the extract of *Cortex angosturæ*. He recovered on the following day. His fellow-traveller remained in a very alarming state for several weeks, but retained sufficient strength of mind to prescribe for himself. His fever was incessant, and complicated with dysentery; but, in his case too, the issue was favourable. At this period no epidemic prevailed in the town, and the air was salubrious; so that the germ of the disease had probably been caught in the damp forests of the Upper Orinoco.

Angostura, so named from its being placed on a narrow part of the river, stands at the foot of a hill of hornblende-slate, destitute of vegetation. The streets are regular, and generally parallel to the course of the stream. The houses are high, agreeable, and built of stone; although the town is not exempt from earthquakes. At the period of this visit the population was only 6000. There is little variety in the surrounding scenery; but the view of the river is singularly majestic. When the waters are high they inundate the quays, and it sometimes happens that even in the streets imprudent persons fall a prey to the crocodiles, which are very numerous.

Humboldt relates that, at the time of his stay at Angostura, an Indian from the island of Margarita having gone to anchor his canoe in a cove where there were not three feet of water, a very fierce crocodile that frequented the spot seized him by the leg and carried him off. With astonishing courage he searched for a knife in his pocket, but not finding it, thrust his fingers into the animal's eyes. The monster, however, did not let go his hold, but plunged to the bottom of the river, and, after drowning his victim, came to the surface and dragged the body to an island.

The number of individuals who perish annually in this manner is very great, especially in villages where the neighbouring grounds are inundated. The same crocodiles remain long in the same places, and become more daring from year to year, especially, as the Indians assert, if they have once tasted human flesh. They are not easily killed, as their skin is impenetrable,—the throat and the space beneath the shoulder being the only parts where a ball or spear can enter. The natives catch them with large iron hooks baited with meat, and attached to a chain fastened to a tree. After the animal has struggled for a considerable time, they attack it with lances.

Affecting examples are related of the intrepidity

of African slaves in attempting to rescue their masters from the jaws of these voracious reptiles. Not many years ago, in the Llanos of Calabozo, a negro, attracted by the cries of his owner, armed himself with a long knife, and, plunging into the river, forced the animal, by scooping out its eyes, to leave its prey and take to flight. The natives being daily exposed to similar dangers think little of them. They observe the manners of the crocodile as the torero studies those of the bull; and quietly calculate the motions of the enemy, its means of attack, and the degree of its audacity.

The general nature of the vast regions bordering on the Orinoco may be sufficiently learned from the above condensed narrative; and we think it unnecessary to follow our learned author through his description of that portion of the river which extends from Angostura to its mouths, especially as it is not founded on personal observation.

CHAPTER XX.

Journey across the Llanos to New Barcelona.

Departure from Angostura—Village of Cari—Natives—New Barcelona—Hot Springs—Crocodiles—Passage to Cumana.

It was night when our travellers for the last time crossed the bed of the Orinoco. They intended to rest near the little fort of San Rafael, and in the morning begin their journey over the Llanos of Venezuela, with the view of proceeding to Cumana or New Barcelona, whence they might sail to the island of Cuba and thence again to Mexico. There they purposed to remain a year, and to take a passage in the galleon from Acapulco to Manilla.

The botanical and geological collections which they had brought from Esmeralda and the Rio Negro had greatly increased their baggage; and as it would have been hazardous to lose sight of such stores, they journeyed but slowly over the deserts, which they crossed in thirteen days. This eastern part of the Llanos, between Angostura and Barcelona, is similar to that already described on the passage from the valley of Aragua to San Fernando de Apure; but the breeze is felt with greater force, although at this period it had ceased. They spent the first night at the house of a Frenchman, a native of Lyons, who received them with the kindest hospitality. He was employed in joining wood by means of a kind of glue called guayca, which resembles the best made from animal substances, and is found between the bark and alburnum of the *Combretum guayca*, a kind of creeping plant.

On the third day they arrived at the missions of Cari. Some showers had recently revived the vegetation. A thick turf was formed of small grasses and herbaceous sensitive plants, while a few fan-palms, rhopalas, and malphighias, rose at great distances from each other. The humid spots were distinguishable by groups of mauritias, which were loaded with enormous clusters of red fruit. The plain undulated from the effect of mirage, the heat was excessive, and the travellers found temporary relief under the shade of the trees, which had, however, attracted numerous birds and insects.

On the 13th July they arrived at the village of Cari, where, as usual, they lodged with the clergyman, who could scarcely comprehend how natives of the north of Europe should have arrived at his dwelling from the frontiers of Brazil. They found more than 500 Caribs in the hamlet, and saw many more at the surrounding missions. They were of large stature, from five feet nine inches to six feet two. The men had the lower part of the body wrapped in a piece of dark-blue cloth, while the women had merely a narrow band. This race differs from the other Indians, not only in being taller, but also in the greater regularity of their features, in having the nose less flattened, and the cheekbones less prominent. The hair of the head is partially shaven, only a circular tuft being left on the top,—a custom that might be supposed to have been borrowed from the monks, but which is equally prevalent among those who have preserved their independence. Both males and females are careful to ornament their persons with paint. The Caribs, once so powerful, now inhabit but a small part of the country which they occupied at the time when America was discovered. They have been exterminated in the West India Islands and the coasts of Darien, but in the provinces of New Barcelona and Spanish Guiana have formed populous villages, under the government of the missions. Humboldt estimates the number inhabiting the Llanos of Piritoo and the banks of the Caroni and Cuyuni at more than 35,000, and the total amount of the pure race at 40,000.

The missionary led the travellers into several huts, where they found the greatest order and cleanliness, but were shocked by the torments that the women inflicted on their infants, for the purpose of raising the flesh in alternate bands from the ankle to the top of the thigh; a practice which the monks had in vain attempted to abolish. This effect was produced by narrow ligatures, which seemed to obstruct the circulation of the blood, although it did not weaken the action of the muscles. The forehead, however, was not flattened, but left in its natural form.

On leaving the mission the philosophers had some difficulty in settling with their Indian muleteers, who had discovered among the baggage the skeletons brought from the cavern of Ataruipe, and were persuaded that the animals which carried such a load would perish on the journey. The Rio Cari was crossed in a boat, and the Rio de Agua Clara by fording. The same objects every where recurred; huts constructed of reeds and roofed with skins; mounted men guarding the herds; cattle, horses, and mules, running half wild. No sheep or goats were seen, these animals being unable to escape from the jaguars.

On the 15th they arrived at the Villa del Pao, where they found some fruit-trees as well as cocoa-palms, which properly belong to the coast. As they advanced the sky became clearer, the soil more dusty, and the atmosphere more fiery. The intense heat, however, was not entirely owing to the temperature of the air, but arose partly from the fine sand mingled with it. On the night of the 16th they rested at the Indian village of Santa Cruz de Cachipo. The warmth had increased so much that they would have preferred travelling by night; but the country was infested by robbers, who murdered the whites that fell into their hands. These were malefactors who had escaped from the prisons on the coast and from the missions, and lived in the Llanos in a manner similar to that of the Bedouin Arabs. Those vast plains, Humboldt thinks, can hardly ever be subjected to cultivation, although he is persuaded that in the lapse of ages, if placed under a government favourable to industry, they will lose much of the wild aspect which they have hitherto retained.

After travelling three days they began to perceive the chain of the mountains of Cumana, which separates the Llanos from the coast of the Caribbean Sea. It appeared at first like a fog-bank, which by degrees condensed, assumed a bluish tint, and became bounded by sinuous outlines. Although the Llanos of Venezuela are bordered on the south by granitic mountains, exhibiting in their broken summits traces of violent convulsions, no blocks were found scattered upon them. The same remark is to be made in regard to the other great plains of South America. These circumstances, as Humboldt remarks, seem to prove that the granitic masses scattered over the sandy plains of the Baltic are a local phenomenon, and must have originated in some great convulsion which took place in the northern regions of Europe.

On the 23d July they arrived at the town of New Barcelona, less fatigued by the heat, to which they had been so long accustomed, than harassed by the sand-wind, that causes painful chaps in the skin. They were kindly received by a wealthy merchant of French extraction, Don Pedro Lavié. This town was founded in 1637, and in 1800 contained more than 16,000 inhabitants. The climate is not so hot as that of Cumana, but very damp, and in the rainy season rather unhealthy. M. Bonpland had by this time regained his strength and activity, but his companion suffered more at Barcelona than he had done at Angostura. One of those extraordinary tropical rains, during which drops of enormous size fall at sunset, had produced uneasy sensations that seemed to threaten an attack of typhus, a disease then prevalent on the coast. They remained nearly a month at Barcelona, where they found their friend Juan Gonzales, who, having resolved to go to Europe, meant to accompany them as far as Cuba.

At the distance of seven miles to the south-east of New Barcelona rises a chain of lofty mountains connected with the Cerro del Bergantin, which is seen from Cumana. When Humboldt's health was sufficiently restored, the travellers made an excursion in that direction, for the purpose of examining the hot-springs in the neighbourhood. These are impregnated with sulphuretted hydrogen, and issue from a quartzose sandstone, lying on a compact limestone resembling that of Jura. The temperature of the water was 109·8°. Their host had lent them one of his finest saddle-horses, warning them at the same time not to ford the little river of Narigual, which is infested with crocodiles. They passed over by a kind of bridge formed of the trunks of trees, and made their animals swim, holding them by the bridles. Humboldt's suddenly disappeared, and the guides conjectured that it had been seized by the caymans.

The crocodiles of the Rio Neveri are numerous, but less ferocious than those of the Orinoco. The people of New Barcelona convey wood to market, by floating the logs on the river, while the proprietors swim here and there to set them loose when they are stopped by the banks. This could not be done in most of the South American rivers infested by those animals. There is no Indian suburb as at Cumana, and the few natives seen in the town, are from the neighbouring missions, or inhabitants of huts scattered in the plain. They are of a mixed race, indolent, and addicted to drinking.

The packet-boats from Corunna to Havannah and Mexico had been due three months, so that they were supposed to have been taken by the English cruisers; when our travellers, anxious to reach Cumana, in order to avail themselves of the first opportunity for Vera Cruz, hired an open vessel. It was laden with cacao, and carried on a contraband trade with the island of Trinidad; for which reason the proprietor thought he had nothing to fear from the British; but they had scarcely reached the narrow channel between the continent and the islands of Borracha and the Chimanas, when they met an armed boat which, hailing them at a great distance, fired some musket-shot at them. It belonged to a privateer of Halifax, and the travellers were forthwith carried on board; but, while Humboldt was negotiating in the cabin, a noise was heard upon deck, and something was whispered to the master, who instantly left him in consternation. An English sloop of war, the Hawk, had come up, and made signals to the latter to bring to; which he not having promptly obeyed, a gun was fired, and a midshipman sent to demand the reason. Humboldt accompanied this officer to the sloop, where Captain Garnier received him with the greatest kindness. Next day they continued their voyage, and at nine in the morning reached the Gulf of Cariaco. The castle of San Antonio, the forest of cactuses, the scattered huts of the Guayquerias, and all the features of a landscape well known to them, rose upon the view; and as they landed at Cumana they were greeted by their numerous friends, who were overjoyed to find untrue a report of their death on the Orinoco, which had been current for several months. The port was every day more strictly blockaded, and the vain expectation of Spanish packets retained them two months and a half longer; during which time they occupied themselves in completing their investigation of the plants of the country; in examining the geology of the eastern part of the peninsula of Araya; and in making astronomical observations, together with experiments on refraction, evaporation, and atmospheric electricity. They also sent off some of their more valuable collections to France.

Having been informed that the Indians brought to the town considerable quantities of native alum found in the mountains, they made an excursion for the purpose of ascertaining its position. Disembarking near Cape Caney they inspected the old salt-pit, now converted into a lake by an irruption of the sea; the ruins of the castle of Araya; and the limestone-mountain of Barigon, which contained fossil shells in perfect preservation. When they visited that peninsula the preceding year, there was a dreadful scarcity of water. But during their absence on the Orinoco it had rained abundantly on various parts along the coast; and the remembrance of these showers occupied the imagination of the natives as a fall of meteoric stones would engage that of the naturalists of Europe.

Their Indian guide was ignorant of the situation of the alum, and they wandered for eight or nine hours among the rocks, which consisted of mica-slate passing into clay-slate, traversed by veins of quartz, and containing small beds of graphite. At length, descending toward the northern coast of the peninsula, they found the substance for which they were searching, in a ravine of very difficult access. Here the mica-slate suddenly changed into carburetted and shining clay-slate, and the springs were impregnated with yellow oxide of iron. The sides of the neighbouring cliffs were covered with capillary crystals of sulphate of alumina, and real beds, two inches thick, of native alum, extended in the clay-slate as far as the eye could reach. The formation appeared to be primitive, as it contained cyanite, rutile, and garnets.

Returning to Cumana, they made preparations for their departure, and availing themselves of an American vessel, laden at New Barcelona for Cuba, they set out on the 16th November, and crossed for the third time the Gulf of Cariaco. The night was cool and delicious, and it was not without emotion that they saw for the last time the disk of the moon illuminating the summits of the cocoa-trees along the banks of the Manzanares. The breeze was strong, and in less than six hours they anchored near the Morro of New Barcelona.

The continental part of the New World is divided between three nations of European origin, of which one, the most powerful, is of Germanic race, and the two others belong to Latin Europe. The latter are more numerous than the former; the inhabitants of Spanish and Portuguese America constituting a population double that of the regions possessed by the English. The French, Dutch, and Danish possessions of the New Continent are of small extent, and the Russian colonies are as yet of little importance. The free Africans of Hayti are the only other people possessed of territory, excepting the native Indians. The British and Portuguese colonists have peopled only the coasts opposite to Europe; but the Spaniards have passed over the Andes, and made settlements in the most western provinces, where alone they discovered traces of ancient civilisation. In the eastern districts, the inhabitants who fell into the hands of the two former nations were wandering tribes or hunters, while in the remoter parts the Spaniards found agricultural states and flourishing empires; and these circumstances have greatly influenced the present condition of these countries. Among other instances may be mentioned the almost total exclusion of African slaves from the latter colonies, and the comfortable condition of the natives of American race, who live by agriculture, and are governed by European laws.

But with respect to the political constitution and relations of the provinces visited by the travellers, it is not expedient here to enter into the details which they have given, more especially as those colonies have lately undergone revolutions that have converted them into independent states, the history of which would afford materials for many volumes. The very interesting sketch of the physical constitution of South America presented by Humboldt must also be passed over, because, in the condensed form to which it would necessarily be reduced, it could not afford an adequate idea of the subject. We must therefore, with our travellers, take leave of Terra Firma, and accompany them on their passage to Havannah.

CHAPTER XXI.

Passage to Havannah, and Residence in Cuba.

Passage from New Barcelona to Havannah—Description of the latter—Extent of Cuba—Geological Constitution—Vegetation—Climate—Population—Agriculture—Exports—Preparations for joining Captain Baudin's Expedition—Journey to Batabano, and Voyage to Trinidad de Cuba.

HUMBOLDT and his companion sailed from the Road of New Barcelona on the 24th November at nine in the evening, and next day at noon reached the island of Tortuga, remarkable for its lowness and want of vegetation. On the 26th there was a dead calm, and about nine in the morning a fine halo formed round the sun, while the temperature of the air fell three degrees. The circle of this meteor, which was one degree in breadth, displayed the most beautiful colours of the rainbow, while its interior and the whole vault of the sky was azure without the least haze. The sea was covered with a bluish scum, which under the microscope appeared to be formed of filaments, that seemed to be fragments of fuci. On the 27th they passed near the island of Orchila, composed of gneiss and covered with plants, and toward sunset discovered the summits of the Roca de Afuera, over which the clouds were accumulated. Indications of stormy weather increased, the waves rose, and waterspouts threatened. On the night of the 2d December a curious optical phenomenon presented itself. The full moon was very high. On its side, forty-five minutes before its passage over the meridian, a great arc suddenly appeared, having the prismatic colours, but of a gloomy aspect. It seemed higher than the moon, had a breadth of nearly two degrees, and remained stationary for several minutes; after which it gradually descended, and sank below the horizon. The sailors were filled with astonishment at this moving arch, which they supposed to announce wind. Next night M. Bonpland and several passengers saw, at the distance of a quarter of a mile, a small flame, which ran on the surface of the sea towards the south-west, and illuminated the atmosphere. On the 4th and 6th they encountered rough weather, with heavy rain accompanied by thunder, and were in considerable danger on the bank of Vibora. At length, on the 19th, they anchored in the port of Havannah, after a boisterous passage of twenty-five days.

Cuba is the largest of the West India Islands, and on account of its great fertility, its naval establishments, the nature of its population,—of which three-fifths are composed of free men,—and its geographical position, is of great political importance. Of all the Spanish colonies it is that which has most prospered; insomuch, that not only has its revenue sufficed for its own wants, but during the struggle between the mother-country and her continental provinces, it furnished considerable sums to the former.

The appearance which Havannah presents at the entrance of the port is exceedingly beautiful and picturesque. The opening is only about 426 yards wide, defended by fortifications; after which a basin, upwards of two miles in its greatest diameter, and communicating with three creeks, expands to the view. The city is built on a promontory, bounded on the north by the fort of La Punta, and on the south by the arsenals. On the western side it is protected by two castles, placed at the distance of 1407 and 2643 yards, the intermediate space being occupied by the suburbs. The public edifices are less remarkable for their beauty than for the solidity of their construction, and the streets are in general narrow and unpaved, in consequence of which they are extremely dirty and disagreeable. But there are two fine public walks to which the inhabitants resort.

Although the town of Havannah, properly so called, is only 1918 yards long and 1066 broad, it contains more than 44,000 inhabitants. The two great suburbs of Jesu-Maria and the Salud accommodate nearly an equal population. In 1810 the amount was as follows:—

```
Whites,..............................................41,227
Free Pardos, or copper-coloured men,... 9,743  ⎫
Free Blacks,................................16,606 ⎭ ...26,349
Pardos Slaves,............................. 2,297  ⎫
Black Slaves,..............................26,431 ⎭ ...28,728
                                                    ───────
                                                    96,304
```

There are two hospitals in the town, the number of sick admitted into which is considerable. Owing to the heat of the climate, the filth of the town, and the influence of the shore, there is usually a great accumulation of disease, and the yellow fever or black vomiting is prevalent. The markets are well supplied.

A peculiar character is given to the landscape in the vicinity of Havannah by the palma real (*Oreodoxa regia*), the trunk of which, enlarged a little towards the middle, attains a height varying from 60 to 85 feet, and is crowned by pinnated leaves rising perpendicularly, and curved at the point. Numerous country-houses of light and elegant construction surround the bay, to which the proprietors retreat when the yellow fever rages in the town.

The island of Cuba is nearly as large as Portugal; its greatest length being 783½ miles, and its mean breadth 51¾ miles. More than four-fifths of its extent is composed of low lands; but it is traversed in various directions by ranges of mountains, the highest of which are said to attain an altitude of 7674 feet. The western part consists of granite, gneiss, and primitive slates; which, as well as the central district, contains two formations of compact limestone, one of argillaceous sandstone, and another of gypsum. The first of these presents large caves near Matanzas and Jaruco, and is filled with numerous species of fossils. The secondary formations to the east of the Havannah are pierced by syenitic and euphotide rocks, accompanied with serpentine. No volcanic eruptions, properly so called, have hitherto been discovered.

Owing to the cavernous structure of the limestone deposites, the great inclination of their strata, the small breadth of the island, and the frequency and nakedness of the plains, there are very few rivers of any magnitude, and a large portion of the territory is subject to severe droughts. Yet the undulating surface of the country, the continually renewed verdure, and the distribution of vegetable forms, give rise to the most varied and beautiful landscapes. The hills and savannahs are decorated by palms of several species, trees of other families, and shrubs constantly covered with flowers. Wild orange-trees ten or fifteen feet in height, and bearing a small fruit, are common, and probably existed before the introduction of the cultivated variety by Europeans. A species of pine (*Pinus occidentalis*) occurs here and in St Domingo, but has not been seen in any of the other West India Islands.

The climate of Havannah, although tropical, is marked by an unequal distribution of heat at different periods of the year, indicating a transition to the climates of the temperate zone. The mean temperature is 78·3°, but in the interior only 73·4°. The hottest months, July and August, do not give a greater average than 82·4°, and the coldest, December and January, present the mean of 69·8°. In summer the thermometer does not rise above 82° or 86°, and its depression in winter so low as 50° or 53·5° is rare. When the north wind blows several weeks, ice is sometimes formed at night at a little distance from the coast, at an inconsiderable elevation above the sea. Yet the great lowerings of temperature which occasionally take place are of so short duration, that the palm-trees, bananas, or the sugar-cane, do not suffer from them. Snow never falls, and hail so rarely that it is only observed during thunder-storms, and with blasts from the S.S.W. once in fifteen or twenty years. The changes however are very rapid, and the inhabitants complain of cold when the thermometer falls quickly to 70°. Hurricanes are of much less frequent occurrence in Cuba than in the other West India Islands.

In 1817 the population was estimated at 630,980. There were 290,021 whites, 115,691 free copper-coloured men, and 225,268 slaves. The original inhabitants have entirely disappeared, as in all the other West India Islands. Intellectual cultivation is almost entirely restricted to the whites; and although in Havannah the first society is not perceptibly inferior to that of the richest commercial cities in Europe, a rudeness of manners prevails in the small towns and plantations.

The common cereal grasses are cultivated in Cuba, together with the tropical productions peculiar to these countries; but the principal exports consist of tobacco, coffee, sugar, and wax. The sugar-cane is planted in the rainy season, from July to October, and cut from February to May. The rapid diminution of wood in the island has caused the want of fuel to be felt in the manufacture of sugar, and Humboldt, during his stay, attempted several new constructions, with the view of diminishing the expenditure of it.*

The tobacco of Cuba is celebrated in every part of Europe. The districts which produce the most aromatic kind are situated to the west of the Havannah, in the Vuelta de Abago; but that grown to the east of the capital on the banks of the Mayari, in the province of Santiago, at Himias, and in other places, is also of excellent quality. In 1827 the produce was about 113,214 cwts., of which 17,888 were exported. The value of this commodity shipped in 1828 was £105,991, 13s. 4d., and in 1829, £142,910. Cotton and indigo, although cultivated, are not to any extent made articles of commerce.

Towards the end of April the travellers, having finished the observations which they had proposed

---

* By the Custom-house returns, 156,158,924 lbs. of sugar were exported from Cuba in 1827; and if the quantity smuggled be estimated at one-fourth more, the total amount would be nearly 200,000,000 lbs. In the same year the exportation of coffee amounted to upwards of 50,000,000 lbs., but it has since fallen off considerably.—See *Macculloch's Dict. of Commerce*, art. Havannah.

to make, were on the point of sailing to Vera Cruz; but intelligence communicated by means of the public papers respecting Captain Baudin's expedition, led them to relinquish the project of crossing Mexico in order to proceed to the Philippine Islands. It had been announced that two French vessels, the Geographe and the Naturaliste, had sailed for Cape Horn, and that they were to go along the coast of Chili and Peru, and from thence to New Holland. Humboldt had promised to join them wherever he could reach the ships, and M. Bonpland resolved to divide their plants into three portions, one of which was sent to Germany by way of England, another to France by Cadiz, and the third left in Cuba. Their friend Fray Juan Gonzales, an estimable young man, who had followed them to the Havannah on his way to Spain, carried part of their collections with him, including the insects found on the Orinoco and Rio Negro; but the vessel in which he embarked foundered in a storm on the coast of Africa. General Don Gonzalo O'Farrill being then in Prussia as minister of the Spanish court, Humboldt was enabled, through the agency of Don Ygnacio, the general's brother, to procure a supply of money; and having made all the necessary preparations for the new enterprise, freighted a Catalonian sloop for Porto Bello, or Carthagena, according as the weather should permit.

On the 6th March the travellers, finding that the vessel was ready to receive them, set out for Batabano, where they arrived on the 8th. This is a poor village surrounded by marshes, covered with rushes and plants of the Iris family, among which appear here and there a few stunted palms.

The marshes are infested by two species of crocodile, one of which has an elongated snout, and is very ferocious. The back is dark-green, the belly white, and the flanks are covered with yellow spots.

On the 9th of March our travellers again set sail in a small sloop, and proceeded through the gulf of Batabano, which is bounded by a low and swampy coast. Humboldt employed himself in examining the influence which the bottom of the sea produces on the temperature of its surface, and in determining the position of some remarkable islands. The water of the gulf was so shallow, that the sloop often struck; but the ground being soft and the weather calm, no damage was sustained. At sunset they anchored near the pass of Don Cristoval, which was entirely deserted, although in the time of Columbus it was possessed by fishermen. The inhabitants of Cuba then employed a singular method for procuring turtles; they fastened a long cord to the tail of a species of *echineis* or sticking-fish, which has a flat disk with a sucking apparatus on its head. By means of this it stuck to the turtle, and was pulled ashore carrying the latter with it. The same artifice is resorted to by the natives of certain parts of the African coast.

They were three days on their passage through the Archipelago of the Jardines and Jardinillos, small islands and shoals partly covered with vegetation; remaining at anchor during the night, and in the day visiting those which were of most easy access. The rocks were found to be fragmentary, consisting of pieces of coral, cemented by carbonate of lime, and interspersed with quartzy sand. On the Cayo Bonito, where they first landed, they observed a layer of sand and broken shells five or six inches thick, covering a formation of madrepore. It was shaded by a forest of rhizophoræ, intermixed with euphorbiæ, grasses, and other plants, together with the magnificent *Tournefortia gnaphalioides*, with silvery leaves and odoriferous flowers. The sailors had been searching for langoustes; but not finding any, avenged themselves on the young pelicans perched on the trees. The old birds hovered around, uttering hoarse and plaintive cries, and the young defended themselves with vigour, although in vain; for the sailors, armed with sticks and cutlasses, made cruel havoc among them. " On our arrival," says Humboldt, " a profound calm prevailed on this little spot of earth; but now every thing seemed to say,—Man has passed here."

On the morning of the 11th they visited the Cayo Flamenco, the centre of which is depressed, and only 15 inches above the surface of the sea. The water was brackish, while in other cayos it is quite fresh; a circumstance difficult to be accounted for in small islands scarcely elevated above the ocean, unless the springs be supposed to come from the neighbouring coast by means of hydrostatic pressure. Humboldt was informed by Don Francisco le Maur, that in the bay of Xagua, to the east of the Jardinillos, fresh water gushes up in several places from the bottom with such force as to prove dangerous for small canoes. Vessels sometimes take in supplies from them; and the lamantins, or fresh-water cetacea, abound in the neighbourhood.

To the east of Cape Flamenco they passed close to the Piedras de Diego Perez, and in the evening landed at Cayo de Piedras, two rocks forming the eastern extremity of the Jardinillos, on which many vessels are lost. They are nearly destitute of shrubs, the shipwrecked crews having cut them down to make signals. Next day, turning round the passage between the northern cape of the Cayo and the island of Cuba, they entered a sea free from breakers, and of a dark-blue colour; the increase of temperature in which indicated a great augmentation of depth. The thermometer was at 79·2°; whereas in the shoal-water of the Jardinillos it had been seen as low as 72·7°, the air being from 77° to 80·6° during the day. Passing in succession the marshy coast of Camareos, the entrance of the Bahia de Xagua, and the mouth of the Rio San Juan, along a naked and desert coast, they entered on the 14th the Rio Guaurabo to land their pilot. Disembarking in the evening, they made preparations for observing the passage of certain stars over the meridian, but were interrupted by some merchants that had dined on board a foreign ship newly arrived, and who invited the strangers to accompany them to the town; which they did, mounted two and two on the same horse. The road to Trinidad is nearly five miles in length, over a level plain covered with a beautiful vegetation, to which the Miraguama palm, a species of corypha, gave a peculiar character. The houses are situated on a steep declivity, about 746 feet above the level of the sea, and command a magnificent view of the ocean, the two ports, a forest of palms, and the mountains of San Juan. The travellers were received with the kindest hospitality by the administrator of the Real Hacienda, M. Munoz. The Teniente Governador, who was nephew to the celebrated astronomer Don Antonio Ulloa, gave them a grand entertainment, at which they met with some French emigrants of Saint Domingo. The evening was passed very agreeably in the house of one of the richest inhabitants, Don Antonio Padron, where they found assembled all the select company of the place. Their departure was very unlike their entrance; for the municipality caused them to be conducted to the mouth of the Rio Guaurabo in a splendid carriage, and an ecclesiastic dressed in velvet celebrated in a sonnet their voyage up the Orinoco.

The population of Trinidad, with the surrounding farms, was stated to be 19,000. It has two ports at the distance of about four miles, Puerto Casilda and Puerto Guaurabo. On their return to the latter of these the travellers were much struck by the prodigious number of phosphorescent insects which illuminated the grass and foliage. These insects (*Elater noctilucus*) are occasionally used for a lamp, being placed in a calabash perforated with holes; and a young woman at Trinidad informed them that, during a long passage from the mainland, she always had recourse to this light when she gave her child the breast at night, the captain not allowing any other on board for fear of pirates.

CHAPTER XXII.

Voyage from Cuba to Carthagena.

Passage from Trinidad of Cuba to Carthagena—Description of the latter—Village of Turbaco—Air-volcanoes—Preparations for ascending the Rio Magdalena.

LEAVING the island of Cuba the travellers proceeded in a S.S.E. direction, and on the morning of the 17th approached the group of the Little Caymans, in the neighbourhood of which they saw numerous turtles of extraordinary size, accompanied by multitudes of sharks. Passing a second time over the great bank of Vibora, they remarked that the colour of the troubled waters upon it was of a dirty-gray, and made observations on the changes of temperature at the surface produced by the varying depth of the sea. On quitting this shoal they sailed between the Baxo Nueva and the lighthouse of Camboy. The weather was remarkably fine, and the surface of the bay was of an indigo-blue or violet tint, on account of the medusæ which covered it. Haloes of small dimensions appeared round the moon. The disappearance of one of them was followed by the formation of a great black cloud, which emitted some drops of rain; but the sky soon resumed its serenity, and a long series of falling-stars and fire-balls were seen moving in a direction contrary to the wind in the lower regions of the atmosphere, which blew from the north. During the whole of the 23d March not a single cloud was seen in the firmament, although the air and the horizon were tinged with a fine red colour; but towards evening large bluish clouds formed, and when they disappeared, converging bands of fleecy vapours were seen at an immense height. On the 24th they entered the kind of gulf bounded by the shores of Santa Martha and Costa Rica, which is frequently agitated by heavy gales. As they advanced toward the coast of Darien the north-east wind increased to a violent degree, and the waves became very rough at night. At sunrise they perceived part of the archipelago of St Bernard, and passing the southern extremity of the Placa de San Bernardo, saw in the distance the mountains of Tigua. The stormy weather and contrary winds induced the master of the vessel to seek shelter in the Rio Sinu, after a passage of sixteen days.

Landing again on the continent of South America, they betook themselves to the village of Zapote, where they found a great number of sailors, all men of colour, who had descended the Rio Sinu in their barks, carrying maize, bananas, poultry, and other articles, to the port of Carthagena. The boats are flat-bottomed, and the wind having blown violently on the coast for ten days, they were unable to proceed on their voyage. These people fatigued the travellers with idle questions about their books and instruments, and tried to frighten them with stories of boas, vipers, and jaguars. Leaving the shores, which are covered with *Rhizophoræ*, they entered a forest remarkable for the great variety of palm-trees which it presented. One of them, the

Æleis melanococca, is only six feet four inches high, but its spathæ contain more than 200,000 flowers, a single specimen furnishing 600,000 at the same time. The kernels of the fruit are peeled in water, and the layer of oil that rises from them, after being purified by boiling, yields the manteca de corozo, which is used for lighting churches and houses.

After an hour's walk they found several inhabitants collecting palm-wine. The tree which affords this liquid is the *Palma dolce* or *Cocos butyracea.* The trunk, which diminishes but little towards the summit, is first cut down, when an excavation eighteen inches long, eight broad, and six in depth, is made below the place at which the leaves and spathæ come off. After three days the cavity is found filled with a yellowish-white juice, having a sweet and vinous flavour, which continues to flow eighteen or twenty days. The last that comes is less sweet, but having a greater quantity of alcohol, it is more highly esteemed. On their way back to the shore they met with Zambos carrying on their shoulders cylinders of palmetto three feet in length, of which an excellent food is prepared. Night surprised them; and, having broken an oar in returning on board, they found some difficulty in reaching the vessel.

The Rio Sinu is of the highest importance for provisioning Carthagena. The gold-washings which were formerly of great value, especially between its source and the village of San Geronimo, have almost entirely ceased, although the province of Antioquia still furnishes, in its auriferous veins, a vast field for mining speculations. It would, however, be of more importance to direct attention to the cultivation of colonial produce in these districts, especially that of cacao, which is of superior quality. The real febrifuge *Cinchona* also grows at the source of the Rio Sinu, as well as in the mountains of Abibé and Maria; and the proximity of the port of Carthagena would enhance its value in the trade with Europe.

On the 27th March the sloop weighed anchor at sunrise. The sea was less agitated, although the wind blew as before. To the north was seen a succession of small conical mountains, rising in the midst of savannahs, where the balsam of Tolu, formerly so celebrated as a medicament, is still gathered. On leaving the gulf of Morosquillo they found the waves swelling so high, that the captain was glad to seek for shelter, and lay to on the north of the village of Rincon; but discovering that they were upon a coral rock, they preferred the open water, and finally anchored near the isle of Arenas, on the night of the 28th. Next day the gale blew with great violence; but they again proceeded, hoping to be able to reach the Boca Chica. The sea was so rough as to break over the deck, and while they were running short tacks, a false manœuvre in setting the sails exposed them for some minutes to imminent danger. It was Palm Sunday; and a Zambo, who had followed them to the Orinoco and remained in their service until they returned to France, did not fail to remind them, that on the same day the preceding year they had undergone a similar danger near the mission of Uruana. After this they took refuge in a creek of the isle of Baru.

As there was to be an eclipse of the moon that night, and next day an occultation of α Virginis,

Humboldt insisted that the captain should allow one of the sailors to accompany him by land to the Boca Chica, the distance being only six miles; but the latter refused, on account of the savage state of the country, in which there was neither path nor habitation; and an incident which occurred justified his prudence. The travellers were going ashore to gather plants by moonlight, when there issued from the thicket a young negro loaded with fetters, and armed with a cutlass. He urged them to disembark on a beach covered with large *Rhizophoræ* among which the sea did not break, and offered to conduct them to the interior of the island of Baru if they would give him some clothes; but his cunning and savage air, his repeated inquiries as to their being Spaniards, and the unintelligible words addressed to his companions who were concealed among the trees, excited their suspicions, and induced them to return on board. These blacks were probably Maroon negroes, who had escaped from prison. The appearance of a naked man, wandering on an uninhabited shore, and unable to rid himself of the chains fastened round his neck and arm, left a painful impression on the travellers; but the sailors felt so little sympathy with these miserable creatures, that they wished to return and seize the fugitives, in order to sell them at Carthagena.

Next morning they doubled the Punta Gigantes, and made sail towards the Boca Chica, the entrance to the port of Carthagena, which is eight or ten miles farther up. On landing, Humboldt learned that the expedition appointed to make a survey of the coast under the command of M. Fidalgo had not yet put to sea, and this circumstance enabled him to ascertain the astronomical position of several places which it was of importance to determine.

During the six days of their stay at Carthagena, they made excursions in the neighbourhood, more especially in the direction of the Boca Grande, and the hill of Popa, which commands the town. The port or bay is nearly eleven miles and a half long. The small island of Tierra Bomba, at its two extremities, which approach, the one to a neck of land from the continent, the other to a cape of the isle of Bani, forms the only entrance to the harbour. One of these, named Boca Grande, has been artificially closed, for the defence of the town, in consequence of an attack attended with partial success made by Admiral Vernon in 1741. The extent of the work was 2640 varas, or 2446 yards, and as the water was from 16 to 20 feet deep, a wall or dike of stone, from 16 to 21 feet high, was raised on piles. The other opening, the Boca Chica, is from 36 to 38 yards broad, but is daily becoming narrower, while the currents acting upon the Boca Grande have opened a breach in it, which they are continually extending.

The insalubrity of Carthagena, which has been exaggerated, varies with the state of the great marshes that surround it. The Cienega de Tesca, which is upwards of eighteen miles in length, communicates with the ocean; and, when in dry years the salt water does not cover the whole plain, the exhalations that rise from it during the heat of the day become extremely pernicious. The hilly ground in the neighbourhood of the town is of limestone, containing petrifactions, and is covered by a gloomy vegetation of cactus, *Jatropha gossypifolia*, croton, and mimosa. While the travellers were searching for plants, their guides showed them a thick bush of *acacia cornigera*, which had acquired celebrity from the following occurrence: A woman, wearied of the well-founded jealousy of her husband, bound him at night with the assistance of her paramour, and threw him into it. The thorns of this species of acacia are exceedingly sharp, and of great length, and the shrub is infested by ants. The more the unfortunate man struggled, the more severely was he lacerated by the prickles, and when his cries at length attracted some persons who were passing, he was found covered with blood, and cruelly tormented by the ants.

At Carthagena the travellers met with several persons whose society was not less agreeable than instructive; and in the house of an officer of artillery, Don Domingo Esquiaqui, found a very curious collection of paintings, models of machinery, and minerals. They had also an opportunity of witnessing the pageant of the Pascua. Nothing, says Humboldt, could rival the oddness of the dresses of the principal personages in these processions. Beggars, carrying a crown of thorns on their heads, asked alms, with crucifixes in their hands, and habited in black robes. Pilate was arrayed in a garb of striped silk, and the apostles, seated round a large table covered with sweetmeats, were carried on the shoulders of Zambos. At sunset, effigies of Jews in French vestments, and formed of straw and other combustibles, were burnt in the principal streets.

Dreading the insalubrity of the town, the travellers retired on the 6th April to the Indian village of Turbaco, situated in a beautiful district, at the entrance of a large forest, about $17\frac{1}{4}$ miles to the south-west of the Popa, one of the most remarkable summits in the neighbourhood of Carthagena. Here they remained until they made the necessary preparations for their voyage on the Rio Magdalena, and for the long journey which they intended to make to Bogota, Popayan, and Quito. The village is about 1151 feet above the level of the sea. Snakes were so numerous that they chased the rats even in the houses, and pursued the bats on the roofs. From the terrace surrounding their habitation, they had a view of the colossal mountains of the Sierra Nevada de Santa Marta, part of which was covered with perennial snow. The intervening space, consisting of hills and plains, was adorned with a luxuriant vegetation, resembling that of the Orinoco. There they found gigantic trees, not previously known, such as the *Rhinocarpus excelsa,* with spirally-curved fruit, the *Ocotea turbacensis,* and the *Cavanillesia platanifolia;* the large five-winged fruit of which is suspended from the tips of the branches like paper lanterns. They botanized every day in the woods from five in the morning till night, though they were excessively annoyed by mosquitoes, zancudoes, xegens, and other tipulary insects. In the midst of these magnificent forests they frequently saw plantations of bananas and maize, to which the Indians are fond of retiring at the end of the rainy season.

The persons who accompanied the travellers on these expeditions often spoke of a marshy ground situated in the midst of a thicket of palms, and which they designated by the name of Los Volcancitos. They said that, according to a tradition preserved in the village, the ground had formerly been ignited, but that a monk had extinguished it by frequent aspersions of holy water, and converted the firevolcano into a water-volcano. Without attaching

Air-volcanoes of Turbaco.

much credit to this tradition, the philosophers desired their guides to lead them to the spot. After traversing a space of about 5300 yards, covered with trunks of *Cavanillesia, Piragra superba,* and *Gyrocarpus,* and in which there appeared here and there projections of a limestone rock containing petrified corals, they reached an open place of about 908 feet square, entirely destitute of vegetation, but margined with tufts of *Bromelia karatas.* The surface was composed of layers of clay of a dark-gray colour, cracked by desiccation into pentagonal and heptagonal prisms. The volcancitos consist of fifteen or twenty small truncated cones rising in the middle of this area, and having a height of from 19 to 25 feet. The most elevated were on the southern side, and their circumference at the base was from 78 to 85 yards. On climbing to the top of these mud-volcanoes, they found them to be terminated by an aperture, from 16 to 30 inches in diameter filled with water, through which air-bubbles obtained a passage; about five explosions usually taking place in two minutes. The force with which the air rises would lead to the supposition of its being subjected to considerable pressure, and a rather loud noise was heard at intervals, preceding the disengagement of it fifteen or eighteen seconds. Each of the bubbles contained from 12 to $14\frac{1}{2}$ cubic inches of elastic fluid, and their power of expansion was often so great that the water was projected beyond the crater, or flowed over its brim. Some of the openings by which air escaped were situated in the plain without being surrounded by any prominence of the ground. It was observed that when the apertures, which are not placed at the summit of the cones, and are enclosed by a little mud wall from 10 to 15 inches high, are nearly contiguous, the explosions did not take place at the same time. It would appear that each crater receives the gas by distinct canals, or that these, terminating in the same reservoir of compressed air, oppose greater or less impediments to the passage of the aeriform fluids. The cones have no doubt been raised by these fluids, and the dull sound that precedes the disengagement of them indicates that the ground is hollow. The natives asserted that there had been no observable change in the form and number of the cones for twenty years, and that the little cavities are filled with water even in the driest seasons. The temperature of this liquid was not higher than that of the atmosphere; the latter having been 81·5°, and the former 80·6° or 81°, at the time of Humboldt's visit. A stick could easily be pushed into the apertures to the depth of six or seven feet, and the dark-coloured clay or mud was exceedingly soft. An ignited body was immediately extinguished on being immersed in the gas collected from the bubbles, which was found to be pure azote.

The stay which our travellers made at Turbaco was uncommonly agreeable, and added greatly to their collection of plants. "Even now," says Humboldt, writing in 1831, "after so long a lapse of time, and after returning from the banks of the Obi and the confines of Chinese Zungaria, these bamboo thickets, that wild luxuriance of vegetation, those orchideæ covering the old trunks of the ocotea and Indian fig, that majestic view of the snowy mountains, that light mist filling the bottom of the valleys at sunrise, those tufts of gigantic trees rising like verdant islets from a sea of vapours, incessantly present themselves to my imagination. At Turbaco we lived a simple and laborious life. We were young; possessed a similarity of taste and disposition; looked forward to the future with hope; were on the eve of a journey which was to lead us to the highest summits of the Andes, and bring us to volcanoes in action in a country continually agitated by earthquakes; and we felt ourselves more happy than at any other period of our distant expedition. The years which have since passed, not all exempt from griefs and pains, have added to the charms of these impressions; and I love to think that, in the midst of his exile in the southern hemisphere, in the solitudes of Paraguay, my unfortunate friend, M. Bonpland, sometimes remembers with delight our botanical excursions at Turbaco, the little spring of Torecillo, the first sight of a gustavia in flower, or of the cavanillesia loaded with fruits having membranous and transparent edges."

M. Bonpland's health having suffered severely during the navigation of the Orinoco and Casiquiare, they resolved to provide themselves with all the conveniences necessary to secure their comfort during the ascent of the Rio Magdalena. They were accompanied on this voyage by an old French physician, M. de Rieux, and two Spaniards. Leaving Turbaco after a stay of ten days, in a cool and very dark night they passed through a wood of bamboos rising from 40 to 50 feet. At daybreak they reached Arjona on the borders of the forest, crossed an arm of the Rio Magdalena in a canoe, and arrived at Mahates, where they had to wait nearly all day for the mules which were to convey their baggage to the place of embarkation. It was excessively hot, without a breath of wind, and to add to their vexation, their only remaining barometer had been broken in passing the canal; but they consoled themselves by examining some beautiful species of parrots which they obtained from the natives.

On the 20th April, at three in the morning, the air feeling deliciously cool, although the thermometer was at 71·6°, they were on their journey to the village of Barancas Nuevas, amid a forest of lofty trees. Half-way between Mahates and that hamlet they found a group of huts elegantly constructed of bamboos, and inhabited by Zambos. Humboldt remarks, that the intermixture of Indians and negroes is very common in those countries, and that the women of the American tribes have a great liking to the men of the African race. To the east of Mahates the limestone formation, containing corals, ceases to appear; the predominant rocks being siliceous with argillaceous cement, forming alternating beds of small-grained quartzose and slaty sandstone, or conglomerates containing angular fragments of lydian-stone, clay-slate, gneiss, and quartz, and varying in colour from yellowish-gray to brownish-red.

Hitherto the narrative of the important journey performed by Humboldt and Bonpland, through those little known but highly interesting regions of South America which were visited by them, has been given as much in detail as is consistent with the nature of a work like the present; but here, as no minute account of their further progress has yet been laid before the public, we must cease to follow them step by step, and content ourselves with a brief narrative of their proceedings.

CHAPTER XXIII.

Brief Account of the Journey from Carthagena to Quito and Mexico.

Ascent of the Rio Magdalena—Santa Fe de Bogota—Cataract of Tequendama—Natural Bridges of Icononzo—Passage of Quindiu—Cargueros—Popayan—Quito—Cotopaxi and Chimborazo—Route from Quito to Lima—Guayaquil—Mexico—Guanaxuato—Volcano of Jorullo—Pyramid of Cholula.

It has been already stated that Humboldt, previously to leaving Paris, had promised Baudin, that should his projected expedition to the southern hemisphere ever take place, he would endeavour to join it; and also that information received by him at Cuba had induced him to relinquish plans subsequently formed, and re-embark for the continent of South America, with the view of proceeding to Guayaquil or Lima, where he expected to meet the navigators. Accordingly he went to Carthagena, where he learned that the season was too far advanced for sailing from Panama to Guayaquil. Giving up, therefore, his intention of crossing the isthmus of Panama, he passed some days in the forests of Turbaco, and afterwards made preparations for ascending the Rio Magdalena.

This river, from its sources near the equator, flows almost directly north. " Nature," says a traveller who sailed up it in 1823, " seems to have designedly dug the bed of the Magdalena in the midst of the cordilleras of Colombia, to form a canal of communication between the mountains and the sea; yet it would have made nothing but an unnavigable torrent, had not its course been stopped in many parts by masses of rock disposed in such a manner as to break its violence. Its waters thus arrested flow gently into the plains of the provinces of Santa Martha and Carthagena, which they fertilize and refresh by their evaporation. Three very distinct temperatures reign on the Magdalena. The sea-breezes blow from its mouth as far as Monpox; from this town to Morales not a breath of air tempers the heat of the atmosphere, and man would become a victim to its power, but for the abundant dews which fall during the night; from Morales as far as the sources of the Magdalena, the south wind moderates the heat of the day, and forms the third temperature. These land-breezes cause the navigation of the Magdalena to be rarely fatal to Europeans."* But, according to the same author, multitudes of animals of various species continually harass the traveller. He cannot bathe on account of the caymans, and if he venture on shore he is in danger of being bitten by serpents.

The voyage up this river, which lasted thirty-five days, was not performed without hazard and inconvenience. Humboldt sketched a chart of it, while his friend was busily occupied in examining the rich and beautiful vegetation of its banks. Disembarking at Honda, they proceeded on mules by dangerous paths, through forests of oaks, melastomæ,

---

* Mollien's Travels in Colombia.

and cinchonæ, to Santa Fe de Bogota the capital of New Grenada. This city stands in a beautiful valley surrounded by lofty mountains, and which would appear to have been at a former period the bed of a great lake. Here the travellers spent several months in exploring the mineralogical and botanical treasures of the country, the magnificent cataract of Tequendama, and the extensive collections of the celebrated Mutis.

The elevated plain on which this metropolis is built, is 8727 feet above the level of the sea, and is consequently higher than the summit of St Bernard. The river of Funza, usually called Rio de Bogota, which drains the valley, has forced its way through the mountains to the south-west of Santa Fe, and near the farm of Tequendama rushes from the plain by a narrow outlet into a crevice, which descends towards the bed of the Rio Magdalena. Respecting this ravine, Gonzalo Ximenes de Quesada, the conqueror of the country, found the following tradition disseminated among the people:— In remote times the inhabitants of Bogota were barbarians, living without religion, laws, or arts. An old man on a certain occasion suddenly appeared among them, of a race unlike that of the natives, and having a long bushy beard. He instructed them in the arts; but he brought with him a very malignant, although very beautiful woman, who thwarted all his benevolent enterprises. By her magical power she swelled the current of the Funza, and inundated the valley; so that most of the inhabitants perished, a few only having found refuge in the neighbouring mountains. The aged visiter then drove his consort from the earth, and she became the moon.

He next broke the rocks that enclosed the valley on the Tequendama side, and by this means drained off the waters; then he introduced the worship of the sun, appointed two chiefs, and finally withdrew to a valley where he lived in the exercise of the most austere penitence during 2000 years.

The cataract of Tequendama presents an assemblage of all that is picturesque. The river a little above it is 144 feet in breadth, but at the crevice narrows to a width of not more than 12 yards. The height of the fall, which forms a double bound, is 574 feet, and the column of vapour that rises from it is visible from Santa Fe at the distance of 17 miles. The vegetation at the foot of the precipice has a totally different appearance from that at the summit; and while the spectator leaves behind him a plain in which the cereal plants of Europe are cultivated, and sees around him oaks, elms, and other trees resembling those of the temperate regions of the northern hemisphere, he looks down upon a country covered with palms, bananas, and sugar-canes.

Leaving Santa Fe, in September 1801, the travellers passed the natural bridges of Iconozo, formed by masses of rock lying across a ravine of immense profundity. The valleys of the cordilleras are generally crevices, the depth of which is often so great, that were Vesuvius seated in them its summit would not exceed that of the nearest mountains. One of these, that, namely, of Iconozo or Pandi, is peculiarly remarkable for the singular form of its rocks, the naked tops of which present the most picturesque contrast with the tufts of trees and shrubs which cover the edges of the gulf. A torrent, named the Summa Paz, forms two beautiful cascades where it enters the chasm, and where it again escapes from it. A natural arch, 47½ feet in length and 39 in breadth, stretches across the fissure at a height of 318 feet above the stream. Sixty-four feet below this bridge is a second composed of three enormous masses of rock, which have fallen so as to support each other. In the middle of it is a hole through which the bottom of the cleft is seen. The torrent, viewed from this place, seemed to flow through a dark cavern, whence arose a doleful sound, emitted by the nocturnal birds that haunt the abyss, thousands of which were seen flying over the surface of the water, supposed by Humboldt from their appearance to be goatsuckers.

In the kingdom of New Grenada, from 2° 30′ to 5° 15′ of north latitude, the cordillera of the Andes is divided into three parallel chains. The eastern one separates the valley of the Rio Magdalena from the plains of the Rio Meta, and on its western declivity are the natural bridges of Icononzo above mentioned. The central chain, which parts the waters between the basin of the Rio Magdalena and that of the Rio Cauca, often attains the limits of perpetual snow, and shoots far beyond it in the colossal summits of Guanacas, Baragan, and Quindiu. The western ridge cuts off the valley of Cauca from the province of Choco and the shores of the South Sea. In passing from Santa Fe to Popayan and the banks of the river now mentioned, the traveller has to descend the eastern chain, either by the Mesa and Tocayma or the bridges of Icononzo, traverse the valley of the Rio Magdalena, and cross the central chain, as Humboldt did, by the mountain of Quindiu.

This mountain, which is considered as the most difficult passage in the cordilleras, presents a thick uninhabited forest, which, in the finest season, cannot be passed in less than ten or twelve days. Travellers usually furnish themselves with a month's provision, as it often happens that the melting of the snow, and the sudden floods arising from it, prevent them from descending. The highest point of the road is $11,499\frac{1}{2}$ feet above the level of the sea, and the path, which is very narrow, has in several places the appearance of a gallery dug in the rock and left open above. The oxen, which are the beasts of burden commonly used in the country, can scarcely force their way through these passages, some of which are 6562 feet in length. The rock is covered with a thick layer of clay, and the numerous gullies formed by the torrents are filled with mud.

In crossing this mountain the philosophers, followed by twelve oxen carrying their collections and instruments, were deluged with rain. Their shoes were torn by the prickles which shoot out from the roots of the bamboos, so that, unwilling to be carried on men's backs, they were obliged to walk barefooted. The usual mode of travelling, however, is in a chair tied to the back of a carguero or porter. When one reflects on the enormous fatigue to which these bearers are exposed, he is at a loss to conceive how the employment should be so eagerly embraced by all the robust young men who live at the foot of the Andes. The passage of Quindiu is not the only part of South America which is traversed in this manner. The whole province of Antioquia is surrounded by mountains so difficult to be crossed, that those who refuse to trust themselves to the skill of a carguero, and are not strong enough to travel on foot, must relinquish all thoughts of leaving the country. The number of persons who follow this laborious occupation, at Choco, Hague, and Medellin, is so great that our travellers sometimes met a file of fifty or sixty. Near the mines of Mexico there are also individuals who have no other employment than that of carrying men on their backs.

The cargueros, in crossing the forests of Quindiu, take with them bundles of the large oval leaves of the vijao, a plant of the banana family, the peculiar varnish of which enables them to resist rain. A hundredweight of these leaves is sufficient to cover a hut large enough to hold six or eight persons. When they come to a convenient spot where they intend to pass the night, the carriers lop a few branches from the trees, with which they construct a frame; it is then divided into squares by the stalks of some climbing plant, or threads of agave, on which are hung the vijao leaves, by means of a cut made in their midrib. In one of these tents, which are cool, commodious, and perfectly dry, our travellers passed several days in the valley of Boquia, amidst violent and incessant rains.

From these mountains, where the truncated cone of Tolima, covered with perennial snow, rises amidst forests of styrax, arborescent passifloræ, bamboos, and waxpalms, they descended into the valley of Cauca towards the west. After resting some time at Cathago and Buga, they coasted the province of Choco, where platina is found among rolled fragments of basalt, greenstone, and fossil wood.

They then went up by Caloto and the mines of Quilichao to Popayan, which is situated at the base of the snowy mountains of Purace and Sotara. This city, the capital of New Grenada, stands in the beautiful valley of the Rio Cauca, at an elevation of 5906 feet above the sea, and enjoys a delicious climate. On the ascent from Popayan towards the summit of the volcano of Purace, at a height of 8694 feet, is a small plain inhabited by Indians, and cultivated with the greatest care. It is bounded by two ravines, on the brink of which is placed a village of the same name. The gardens, which are enclosed with hedges of euphorbium, are watered by the springs that issue abundantly from the porphyritic rock; and nothing can be more agreeable than the contrast between the beautiful verdure of this plain and the chain of dark mountains surrounding the volcano. The hamlet of Purace, which the travellers visited in November 1801, is celebrated for the fine cataracts of the Rio Vinagre, the waters of which are acid. This little river is warm towards its source, and after forming three falls, one of which is 394 feet in height and is exceedingly picturesque, joins the Rio Cauca, which for 14 miles below the junction is destitute of fish. The crater of the volcano is filled with boiling water, which, amid frightful noises, emits vapours of sulphuretted hydrogen.

The travellers then crossed the precipitous cordilleras of Almaquer to Pasto, avoiding the infected and contagious atmosphere of the valley of Patia. From the latter town, which is situated at the foot of a burning volcano, they traversed the elevated platform of the province of Los Pastos, celebrated for its great fertility; and after a journey of four months, performed on mules, arrived at Quito on the 6th January 1802.

The climate of this province is remarkably agreeable, and almost invariable. During the months of December, January, February, and March, it generally rains every afternoon from half-past one to five; but even at this season the evenings and mornings are most beautiful. The temperature is so mild that vegetation never ceases. "From the terrace of the government palace there is one of the most enchanting prospects that human eye ever witnessed, or nature ever exhibited. Looking to the south, and glancing along towards the north, eleven mountains covered with perpetual snow present themselves, their bases apparently resting on the verdant hills that surround the city, and their heads piercing the blue arch of heaven, while the clouds hover midway down them, or seem to crouch at their feet. Among these the most lofty are Cayambeurcu, Imbaburu, Ilinisa, Antisana, Chimborazo, and the beautifully-magnificent Cotopaxi, crowned with its volcano."*

Nearly nine months were devoted to researches of various kinds. They made excursions to the snowy mountains of Antisana, Cotopaxi, Tunguragua, and Chimborazo, the latter of which was considered as the highest on the globe until it was found to be exceeded by some of the colossal summits of the Himmaleh, and even by several in Upper Peru. In all these journeys they were accompanied by a young man, son of the Marquis of Selva-alègre, who subsequently followed them to Peru and Mexico.† They twice ascended to the

---

* Stevenson's Residence in South America, vol. ii. p. 324.
† This accomplished individual, Don Carlos Montufar, of whom our author speaks with approbation, having connected himself with the popular party in the struggles of which the Spanish colonies have lately been the theatre, was seized in Quito, in 1811, by Don

volcanic summit of Pichincha, where they made experiments on the constitution of the air,—its elasticity, its electrical, magnetic, and hygroscopic qualities,—and the temperature of boiling water.

Cotopaxi is the loftiest of those volcanoes of the Andes which have produced eruptions at recent periods; its absolute height being 18,878 feet. It is consequently 2625 feet higher than Vesuvius would be were it placed on the top of the Peak of Teneriffe. The scoriæ and rocks ejected by it, and scattered over the neighbouring valleys, would form a vast mountain of themselves. In 1738 its flames rose 2953 feet above the crater; and in 1744 its roarings were heard as far as Honda, on the Magdalena, at a distance of 690 miles. On the 4th April 1768, the quantity of ashes thrown out was so great, that in the towns of Hambato and Tacunga the inhabitants were obliged to use lanterns in the streets. The explosion which took place in January 1803 was preceded by the sudden melting of the snows which covered the surface; and our travellers, at the port of Guayaquil, $179\frac{1}{2}$ miles distant, heard day and night the noises proceeding from it, like discharges of a battery.

This celebrated mountain is situated to the southeast of Quito, at the distance of 41 miles, in the midst of the Andes. Its form is the most beautiful and regular of all the colossal summits of that mighty chain; being a perfect cone, which is covered with snow, and shines with dazzling splendour at sunset. No rocks project through the icy covering,

---

Toribio Montes, sentenced as a traitor, and shot through the back; after which his heart was taken out and burnt.—See *Stevenson's Residence in South America*, vol. iii. p. 44.

except near the edge of the crater, which is surrounded by a small circular wall. In ascending it is extremely difficult to reach the lower boundary of the snows, the cone being surrounded by deep ravines; and, after a near examination of the summit, Humboldt thinks he may assert that it would be altogether impossible to reach the brink of the crater.

It was mentioned that, in the kingdom of New Grenada, the cordilleras of the Andes form three chains, in the great longitudinal valleys of which flow two large rivers. To the south of Popayan, on the table-land of Los Pastos, these three chains unite into a single group, which stretches far beyond the equator. This group, in the kingdom of Quito, presents an extraordinary appearance from the river of Chota, the most elevated summits being arranged in two lines, forming as it were a double ridge to the cordilleras. These summits served for signals to the French academicians when employed in the measurement of an equinoctial degree. Bouguer considered them as two chains, separated by a longitudinal valley; but this valley Humboldt views as the ridge of the Andes itself. It is an elevated plain, from 8858 to 9515 feet above the level of the sea; and the volcanic summits of Pichincha, Cayambo, Cotopaxi, and other celebrated peaks, are, he thinks, so many protuberances of the great mass of the Andes. In consequence of the elevation of the territory of Quito, these mountains do not seem so high as many of much inferior altitude rising from a lower basis.

On Chimborazo the line marking the inferior limit of perpetual snow is at a height somewhat exceeding that of Mont Blanc. On a narrow ledge, which rises amidst the snows on the southern declivity, our travellers attempted on the 23d June to reach the summit. The point where they stopped to observe the inclination of the magnetic meridian was more elevated than any yet attained by man, being 3609 feet higher than the summit of Mont Blanc, and more than 3714 feet higher than La Condamine and Bouguer reached in 1745 on the Corazon. The ridge to which they climbed, and beyond which they were prevented from proceeding by a deep chasm in the snow, was 19,798 feet above the level of the sea; but the summit of the mountain was still 1439 feet higher. The blood issued from their eyes, lips, and gums. The form of Chimborazo is conical, but the top is not truncated like that of Cotopaxi, being rounded or semicircular in outline.

While at Quito, Humboldt received a letter from the National Institute of France, by which he was apprized that Captain Baudin had set out for New Holland by the Cape of Good Hope. He was obliged therefore to renounce all thoughts of joining the expedition, although the hope of being able to meet it had induced him to relinquish his plan of proceeding from Cuba to Mexico and the Philippine Islands, and had led him upwards of 3452 miles southward. The travellers, however, consoled themselves with the thought of having examined regions over which the eye of science had never before glanced; and, resolving henceforth to trust solely to their own resources, after spending some months in exploring the Andes, they set out in the direction of Lima.

They first pointed their course to the great River

Amazon, visiting the ruins of Lactacunga, Hambato, and Riobamba, in a country the face of which was entirely changed by the frightful earthquakes of 1797, that destroyed nearly 40,000 of the inhabitants. They then with great difficulty passed to Loxa, where in the forests of Gonzanama and Malacates they examined the trees which yield the Peruvian bark. The vast extent of ground which they traversed in the course of their expedition afforded them better opportunities than any botanist had ever enjoyed of comparing the different species of Cinchona.

Leaving Loxa they entered Peru by Ayavaca and Gouncabamba, traversing the ridge of the Andes to descend to the River Amazon. In two days they had to cross thirty-five times the Rio de Chayma. They saw the magnificent remains of the causeway of the Incas, which traversed the porphyritic summits from Cusco to Assouay, at a height varying from 7670 to 11,510 feet. At the village of Chamaya, on a river of the same name, they took ship and descended to the Amazon.

La Condamine, on his return from Quito to Para, embarked on this river only below Quebrada de Chuchunga; and Humboldt, with the view of completing the map made by the French astronomer, proceeded as far as the cataracts of Rentama. At Tomependa, the principal place of the province of Jaen de Bracamorros, he constructed a map of the Upper Amazon, from his own observations as well as from accounts received from the natives. Bonpland employed himself, as usual, in examining the subjects of the vegetable kingdom, among which he discovered several new species of Cinchona.

Returning to Peru, our travellers crossed the cordillera of the Andes the fifth time. In seven degrees of south latitude they determined the position of the magnetic equator, or the line in which the needle has no inclination. They also examined the mines of Hualgayoc, where large masses of native silver are found at an elevation of 12,790 feet above the sea, and which, together with those of Pasco and Huantajayo, are the richest in Peru. From Caxamarca, celebrated for its hot-springs and the ruins of the palace of Atahualpa, they went down to Truxillo. In this neighbourhood are the remains of the ancient Peruvian city Mansiche adorned by pyramids, in one of which an immense quantity of gold was discovered in the eighteenth century. Descending the western slope of the Andes they beheld for the first time the Pacific Ocean, and the long narrow valley bounded by its shores, in which rain and thunder are unknown. From Truxillo they followed the arid coast of the South Sea, and arrived at Lima, where they remained several months. At the port of Callao, Humboldt had the satisfaction of observing the transit of Mercury, although the thick fog which prevails there sometimes obscures the sun for many days in succession.

In January 1803 the travellers embarked for Guayaquil, in the vicinity of which they found a splendid forest of palms, plumeriæ, tabernæ-montanæ, and scitamineæ. Here also they heard the incessant noises of the volcano of Cotopaxi, which had experienced a tremendous agitation on the 6th January. From Guayaquil they proceeded by sea to Acapulco in New Spain. At first, Humboldt's intention was to remain only a few months in Mexico, and return as speedily as possible to Europe, more especially as his instruments, and in particular the chronometers, were getting out of order, while he found it impossible to procure others. But the attractions of so beautiful and diversified a country, the great hospitality of its inhabitants, and the dread of the yellow fever of Vera Cruz, which usually attacks those who descend from the mountains between June and October, induced him to remain until the middle of winter.

After making numerous observations and experiments on the atmospherical phenomena, the horary variations of the barometer, magnetism, and the natural productions of the country, our travellers set out in the direction of Mexico; gradually ascending by the burning valleys of Mescala and Papagayo, where the thermometer rose to 89·6° in the shade, and where the river is crossed on fruits of *Crescentia pinnata*, attached to each other by ropes of agave. Reaching the elevated plains of Chilpantzuigo, Tehuilotepec, and Tasco, which are situated at a height varying from 3837 to 4476 feet above the sea, they entered a region blessed with a temperate climate, and producing oaks, cypresses, pines, tree-ferns, and the cultivated cereal plants of Europe. After visiting the silver-mines of Tasco, the oldest and formerly the richest of Mexico, they went up by Cuernaraca and Guachilaco to the capital. Here they spent some time in the agreeable occupation of examining numerous curiosities, antiquities, and institutions, in making astronomical observations, in studying the natural productions of the surrounding country, and in enjoying the society of enlightened individuals. The longitude of Mexico, which had been misplaced two degrees on the latest maps, was accurately determined by a long series of observations.

Our travellers next visited the celebrated mines of Moran and Real del Monte, and examined the obsidians of Oyamel, which form layers in pearlstone and porphyry, and were employed by the ancient Mexicans for the manufacture of knives. The cascade of Regla, a representation of which forms the vignette to the present volume, is situated in the neighbourhood. The regularity of the basaltic columns is as remarkable as that of the deposites of Staffa. Most of them are perpendicular; though some are horizontal, and others have various degrees of inclination. They rest upon a bed of clay, beneath which basalt again occurs. Returning from this excursion in July 1803, they made another to the northern part of the kingdom, in the course of which they inspected the aperture made in the mountain of Suicog for the purpose of draining the valley of Mexico. They next passed by Queretaro, Salamanca, and the fertile plains of Yrapuato, on the way to Guanaxuato, a large city placed in a narrow defile, and celebrated for its mines.

There they remained two months, making researches into the geology and botany of the neighbouring country. From thence they proceeded by the valley of San Jago to Valladolid, the capital of the ancient kingdom of Mechoacan; and, notwithstanding a continuance of heavy autumnal rains, descended by Patzquaro, which is situated on the edge of an extensive lake towards the shores of the Pacific Ocean, to the plains of Jorullo. Here they entered the great crater, making their way over crevices exhaling ignited sulphuretted hydrogen, and experiencing much danger from the brittleness of the lava.

The formation of this volcano is one of the most

extraordinary phenomena which have been observed on our globe. The plain of Malpais, covered with small cones from six to ten feet in height, is part of an elevated table-land bounded by hills of basalt, trachyte, and volcanic tufa. From the period of the discovery of America to the middle of the last century, this district had undergone no change of surface, and the seat of the crater was then covered with a plantation of indigo and sugar-cane; when, in June 1759, hollow sounds were heard, and a succession of earthquakes continued for two months, to the great consternation of the inhabitants. From the beginning of September every thing seemed to announce the re-establishment of tranquillity; but in the night of the 28th the frightful subterranean noises again commenced. The Indians fled to the neighbouring mountains. A tract not less than from three to four square miles in extent rose up in the shape of a dome; and those who witnessed the phenomenon asserted, that flames were seen issuing from a space of more than six square miles, while fragments of burning rocks were projected to an immense height, and the surface of the ground undulated like an agitated sea. Two brooks which watered the plantations precipitated themselves into the burning chasms. Thousands of the small cones described above, suddenly appeared, and in the midst of these eminences, called hornitos or ovens, six great masses, having an elevation of from 1312 to 1640 feet above the original level of the plain, sprung up from a gulf running from N.N.E. to S.S.W. The most elevated of these mounds is the great volcano of Jorullo, which is continually burning. The eruptions of this central volcano continued till February 1760, when they became less frequent. The Indians, who had abandoned all the villages within thirty miles of it, returned once more to their cottages, and advanced towards the mountains of Aguasarco and Santa Ines, to contemplate the streams of fire that issued from the numberless apertures. The roofs of the houses of Queretaro, more than 166 miles distant, were covered with volcanic dust. Mr Lyell (Principles of Geology, vol. i. p. 379) states, on the authority of Captain Vetch, that another eruption happened in 1819, accompanied by an earthquake, during which ashes fell at the city of Guanaxuato, 140 miles distant from Jorullo, in such quantities as to lie six inches deep in the streets.

When Humboldt visited this place, the natives assured him that the heat of the hornitos had formerly been much greater. The thermometer rose to 203° when placed in the fissures exhaling aqueous vapour. Each of the cones emitted a thick smoke, and in many of them a subterranean noise was heard, which seemed to indicate the proximity of a fluid in ebullition. Two streams were at that period seen bursting through the argillaceous vaults, and were found by the traveller to have a temperature of 126·9°. The Indians give them the names of the two rivers which had been engulfed, because in several parts of the Malpais great masses of water are heard flowing in a direction from east to west. Our author considers all the district to be hollow; but Scrope and Lyell find it more suitable to their views of volcanic agency to represent the conical form of the ground as resulting from the flow of lava over the original surface of the plain.

The Indians of this province are represented as being the most industrious of New Spain. They

Costumes of the Indians of Mechoacan.

have a remarkable talent for cutting out images in wood, and dressing them in clothes made of the pith of an aquatic plant, which being very porous imbibes the most vivid colours. Two figures of this kind, which Humboldt brought home for the Queen of Prussia, are here represented. They exhibit the characteristic traits of the American race, together with a strange mixture of the ancient costume with that which was introduced by the Spaniards.

From Valladolid, the ancient kingdom of Mechoacan, the travellers returned to Mexico by the elevated plain of Tolucca, after examining the volcanic mountains in the vicinity. They also visited the celebrated cheiranthostæmon of Cervantes, a tree of which it was at one time supposed there did not exist more than a single specimen.

At that city they remained several months, for the

purpose of arranging their botanical and geological collections, calculating the barometrical and trigonometrical measurements which they had made, and sketching the plates of the Geological Atlas which Humboldt proposed to publish. They also assisted in placing a colossal equestrian statue of the king, which had been cast by a native artist. In January 1804 they left Mexico with the intention of examining the eastern declivity of the cordillera of New Spain. They also measured the great pyramid of Cholula, an extraordinary monument of the Toltecks, from the summit of which there is a splendid view of the snowy mountains and beautiful plains of Tlascala. It is built of bricks, which seemed to have been dried in the sun, alternating with layers of clay. They then descended to Xalapa, a city placed at an elevation of 4138 feet above the sea, in a delightful climate. The dangerous road which leads from it to Perote, through almost impenetrable forests, was thrice barometrically levelled by Humboldt. Near the latter place is a mountain of basaltic porphyry, remarkable for the singular form of a small rock placed on its summit, and which is named the Coffer of Perote. This elevation commands a very extensive prospect over the plain of Puebla and the eastern slope of the cordilleras of Mexico, which is covered with dense forests. From it they also saw the harbour of Vera Cruz, the castle of St Juan of Ulloa, and the seacoast.

Before following our travellers across the Atlantic, it may be useful to present a sketch of the valuable observations recorded in Humboldt's Political Essay on the Kingdom of New Spain, and which are in part the result of his researches in that interesting country.

CHAPTER XXIV.

Description of New Spain or Mexico.

General Description of New Spain or Mexico—Cordilleras—Climates—Mines—Rivers—Lakes—Soil—Volcanoes—Harbours—Population—Provinces—Valley of Mexico, and Description of the Capital—Inundations, and Works undertaken for the Purpose of preventing them.

PREVIOUS to Humboldt's visit to New Spain, the information possessed in Europe respecting that interesting and important country was exceedingly meagre and incorrect. The ignorance of the European conquerors, the indolence of their successors, the narrow policy of the government, and the want of scientific enterprise among the Creoles and Spaniards, left it for centuries a region of dim obscurity into which the eye of research was unable to penetrate. So inaccurate were the maps, that even the latitude and longitude of the capital remained unfixed, and the inhabitants were thrown into consternation by the occurrence of a total eclipse of the sun on the 21st February 1803; the almanacs, calculating from a false indication of the meridian, having announced it as scarcely visible. The determination of the geographical position of many of the more remarkable places, that of the altitude of the volcanic summits and other eminences, together with the vast mass of intelligence contained in the Political Essay on New Spain, served to dispel in some measure the darkness; and since the period of Humboldt's visit numerous travellers have contributed so materially to our acquaintance with Mexico, that it no longer remains among the least known of those remote countries of the globe over which the power of Europe has extended.

Although the independence of the American states has now been confirmed, and their political relations entirely changed since the time our author was there, the aspect of nature continues the same in those extensive regions; and, as we have less to do with their history and national circumstances than with the discoveries of the learned traveller, we shall follow, as heretofore, his descriptions of the countries examined by him in the relations in which they then stood.

The Spanish settlements in the New Continent formerly occupied that immense territory comprised between 41° 43' of south latitude and 37° 48' of north latitude, equalling the whole length of Africa, and exceeding the vast regions possessed by the Russian empire or Great Britain in Asia. They were divided into nine great governments, of which five, viz. the viceroyalties of Peru and New Grenada, the capitanias-generales of Guatimala, Porto Rico, and Caraccas, are entirely intertropical, while the other four, viz. the viceroyalties of Mexico and Buenos Ayres, and the capitanias-generales of Chili and Havannah, including the Floridas, are chiefly situated in the temperate zones. Mexico was the most important, as well as the most civilized of the whole, and was long considered as such by the court of Madrid.

The name of New Spain was at first given in 1518 to the province of Yucatan, where the companions of Grijalva were astonished at the civilisation of the inhabitants. Cortez employed it to denote the whole empire of Montezuma, though it was subsequently used in various senses. Humboldt designates by it the vast country which has for its northern and southern limits the parallels of 38° and 16°. The length of this region from S.S.E. to N.N.W. is nearly 1678 miles; its greatest breadth 994 miles. The isthmus of Tehuantepec, to the south-east of the port of Vera Cruz, is the narrowest part; the distance from the Atlantic Ocean to the South Sea being there only 155 miles. The question of opening a communication by a canal between the two oceans at this point, the isthmus of Panama, or several others which he mentions, is fully discussed by the author. He discredits the idea that the level of the South Sea is higher than that of the Gulf of Mexico, and imagines that were a rupture of the intervening barrier effected, the current would establish itself in the direction opposite to that usually apprehended.

When a general view is taken of the whole surface of Mexico, it is seen that one-half is situated within the tropic, while the rest belongs to the temperate zone. This latter portion contains 775,019 square miles. The physical climate of a country does not altogether depend upon its distance from the pole, but also upon its elevation, its proximity to the ocean, and other circumstances; so that of the 645,850 square miles in the torrid zone, more than three-fifths have a cold, or at least temperate atmosphere. The whole interior of Mexico, in fact, constitutes an immense table-land, having an elevation which varies from 6562 to 8202 feet above the level of the sea.

The chain of mountains which forms this vast plain is continuous with the Andes of South America. In the southern hemisphere the cordillera is every where broken up by fissures or valleys of small breadth; but in Mexico it is the ridge itself that constitutes the platform. In Peru the most elevated summits form the crest of the Andes, while in the other the prominences are irregularly scattered over the plain, and have no relation of parallelism to the direction of the cordillera. In Peru and New Grenada there are transverse valleys, having sometimes 4590 feet of perpendicular depth, which entirely prevent the use of carriages; while in New Spain vehicles are used along an extent of more than 1726 miles. The general height of the table-land of Mexico is equal to that of Mount Cenis, St Gothard; or the Great St Bernard of the Swiss Alps; and to determine this circumstance Humboldt executed five laborious barometrical surveys, which enabled him to construct a series of vertical sections of the country.

In South America the cordillera of the Andes presents plains completely level at immense altitudes, such as that on which the city of Santa Fe de Bogota stands, that of Caxamarca in Peru, and those of Antisana, which exceed in height the summit of the Peak of Teneriffe. But all these levels are of small extent, and being separated by deep valleys are of difficult access. In Mexico, on the other hand, vast tracts of champaign country are so approximated to each other as to form but a single plain occupying the elongated ridge of the cordil-

DIVERSITY OF CLIMATE. 347

lera, and running from the 18th to the 40th degree of north latitude. The descent towards the coasts is by a graduated series of terraces, which oppose great difficulties to the communication between the maritime districts and the interior, presenting at the same time an extraordinary diversity of vegetation.

The plains along the coasts are the only parts that possess a climate adapted to the productions of the West Indies,—the mean temperature of those situated within the tropics, and whose elevation does not exceed 984 feet, being from 77° to 78·8°, which is several degrees greater than the mean temperature of Naples. These fertile regions, which produce sugar, indigo, cotton, and bananas, are named *Tierras calientes*. Europeans remaining in them for any considerable time, particularly in the towns, are liable to the yellow fever or black vomiting. On the eastern shores the great heats are occasionally tempered by strata of refrigerated air brought from the north by the impetuous winds that blow from October to March, which frequently cool the atmosphere to such a degree, that at Havannah the thermometer descends to 32°, and at Vera Cruz to 60·8°.

On the declivities of the cordillera, at the elevation of 3937 or 4921 feet, there prevails a mild climate, never varying more than four or five degrees. To this region, of which the mean annual temperature is from 68° to 69·8°, the natives give the name of *Tierras templadas*. Unfortunately these tracts are frequently covered with thick fogs, as they occupy the height to which the clouds usually ascend above the level of the sea.

The plains which are elevated more than 7218 feet above that level, and of which the mean temperature is under 62·6°, are named *Tierras frias.* The whole table-land of Mexico belongs to this description, which the natives consider cold, although the ordinary warmth is equal to that of Rome. There are plains of still greater elevation, on which, although they have a mean temperature of from 51·8° to 55·4°, equal to that of France and Lombardy, the vegetation is less vigorous, and European plants do not thrive so well as in their native soil. The winters there are not extremely severe, but in summer the sun has not sufficient power over the rarified air to bring fruits to perfect maturity.

From the peculiar circumstances of New Spain, as here sketched, the influence of geographical position upon the vegetation is much less than that of the height of the ground above the sea. In the nineteenth and twentieth degrees of latitude, sugar, cotton, cacao, and indigo, are produced abundantly only at an elevation of from 1968 to 2625 feet. Wheat thrives on the declivities of the mountains, along a zone which commences at 4593 feet and ends at 9843. The banana (*Musa paradisiaca*), on the fruit of which the inhabitants of the tropics chiefly subsist, is seldom productive above 5085 feet; oaks grow only between 2625 and 9843 feet; and pines never descend lower than 6096, nor rise above 13,124 feet.

The internal provinces of the temperate zone enjoy a climate essentially different from that of the same parallels in the Old Continent. So remarkable an inequality prevails indeed between the temperature of the seasons, that while the winters resemble those of Germany the summers are like those of Sicily. A similar difference exists between

MINES—RIVERS—LAKES. 349

the other parts of America and the corresponding latitudes in Europe; but it is less perceptible on the western than on the eastern coasts.

New Spain possesses a peculiar advantage in the circumstances under which the precious metals have been deposited. In Peru the most important silver mines, those of Potosi, Pasco, and Chota, are placed at an immense elevation; so that, in working them, men, provisions, and cattle, must be brought from a distance; but in Mexico the richest of these, those, namely, of Guanaxuato, Zacatecas, Tasco, and Real del Monte, are at moderate heights, and surrounded by cultivated fields, towns, and villages.

There are few rivers of consequence in the country, the Rio Bravo del Norte and the Rio Colorado being the only ones of any magnitude. The former has a course of 1767 miles, the latter of 863; but these streams flow in the least cultivated parts of the country, and can have little influence in a commercial point of view until colonization shall extend to their shores. In the whole equinoctial part of New Spain there are only small rivulets, of which very few can ever become interesting to the merchant.

The numerous lakes, the greater part of which appear to be annually decreasing in size, are the remains of immense basins of water that formerly existed on the elevated plains. Of these may be mentioned the lake of Chapala, nearly 2067 square miles in extent; those of the valley of Mexico, which comprehend a fourth part of its surface; that of Patzcuaro in Valladolid; and, finally, the lakes of Mexitlan and Parras in New Biscay.

The interior of New Spain, and especially a great part of the elevated table-land of Anahuac, is arid and destitute of vegetation; which arises from the rapid evaporation in high plains, and the circumstance that few of the mountains enter the region of perpetual snow, which under the equator commences at the height of 15,748 feet, and in the 45th degree of latitude at that of 8366 feet. In Mexico, in the 19th and 20th degrees, perpetual frost commences, according to Humboldt's measurements, at 15,092 feet of elevation; so that of the six colossal summits, which are placed in the same line in the 19th parallel of latitude, only four, namely, the Peak of Orizaba, Popocatepetl, Iztaccihuatl, and Nevado de Tolucca, are clothed with perennial snow; while the Cofre de Perote and the Volcan de Colmia remain uncovered during the greater part of the year. None of the other mountains rise into so lofty a region.

In general, in the equinoctial part of New Spain, the soil, climate, and vegetation, present a similar character to those of the temperate zone. Although the table-lands are singularly cold in winter, the temperature is much higher in summer than in the Andes of Peru, because the great mass of the cordillera of Mexico, and the vast extent of its plains, produce a reverberation of the sun's rays never observed in elevated countries of greater inequality.

To the north of 20° the rains, which fall only in June, July, August, and September, very seldom extend to the interior. The mountains, being composed of porous amygdaloid and fissured porphyries, present few springs; the filtrated water losing itself in the crevices opened by ancient volcanic eruptions, and issuing at the bottom of the cordilleras.

The aridity of the central plain, on which there

is a great deficiency of wood, is prejudicial to the working of the mines; and this natural evil has been augmented since the arrival of Europeans, who have not only destroyed the trees without planting others, but have drained a large extent of ground, and thus increased the saline efflorescences which cover the surface and are hostile to cultivation. This dryness, however, is confined to the more elevated plains; and the declivities of the cordillera being exposed to humid winds and fogs, their vegetation is uncommonly vigorous.

Mexico is less disturbed by earthquakes than Quito, Guatimala, and Cumana, although these destructive commotions are by no means rare on the western coasts, and in the neighbourhood of the capital, where, however, they are never so violent as in other parts of America. There are only five active volcanoes in all New Spain: Orizaba, Popocatepetl, Tustla, Jorullo, and Colima.

The physical situation of that kingdom confers inestimable advantages upon it, in a commercial point of view. Under careful cultivation it is capable of producing all that commerce brings together from every part of the globe: sugar, cochineal, cacao, cotton, coffee, wheat, hemp, flax, silk, oil, and wine. It furnishes every metal, not even excepting mercury, and is supplied with the finest timber; but the coasts oppose obstacles which it will be difficult to overcome. The western shores are indeed furnished with excellent harbours; but the eastern are almost entirely destitute of them, the mouths of the rivers there being choked up with sands, which are constantly adding to the land. Vera Cruz, the principal port on this side, is merely an open road. Both coasts, too, are rendered inaccessible for several months by severe tempests, which prevent all navigation. The north winds, *los 'nortes,* prevail in the Mexican Gulf from the autumnal to the vernal equinox. They are very violent in March, though usually more moderate in September and October. The navigators who have long frequented the port of Vera Cruz are familiar with the symptoms of the coming storm, which is preceded by a great change in the barometer, and a sudden interruption in the regular occurrence of its horary oscillations. At first a gentle land-wind blows from W.N.W., and is succeeded by a breeze rising from the N.E. then from the S. A suffocating heat succeeds, and the water dissolved in the atmosphere is precipitated on the walls and pavements. The summits of Orizaba, of the Cofre de Perote, and the mountains of Villa Rica, are cloudless, while their bases are concealed by vapours. In this state of the air the tempest commences, usually with great impetuosity, and generally continues three or four days. Occasionally, even in May, June, July, and August, violent hurricanes are experienced in the Gulf of Mexico. The navigation of the western coasts is very dangerous in July and August, when sudden gales burst from the S.W.; and even in the fine season, from October to May, furious winds sometimes blow from the N.E. and N.N.E. In short, all the coasts of New Spain are at certain periods dangerous to navigators.

It is probable that Mexico was formerly better inhabited than it is at present; but its population was concentrated in a very small space in the neighbourhood of the capital. At the present day it is more generally distributed than it was before the conquest, and the number of Indians has increased during the last century. According to an imperfect census made in 1794, the return was estimated at 5,200,000. The proportion of births to deaths, during the time between that period and Humboldt's visit, was found, from data furnished by the clergy, to be 170 : 100; while that of births to the total amount he considers as 1 in 17, and of the deaths as 1 in 30. The annual number at present born he estimates at nearly 350,000, and that of deaths at 200,000. It would thus appear that, if this rate of increase were not checked from time to time by some extraordinary cause, the population of New Spain would double every nineteen years. In the United States generally it has doubled, since 1784, every twenty or twenty-three years; and in some of them it doubles in thirteen or fourteen. In France, on the other hand, the number of inhabitants would double in 214 years were no wars or contagious diseases to interfere. Such is the difference between countries that have long been densely peopled and those whose civilisation is of recent date. Humboldt, from various considerations, assumes the population of Mexico in 1803 at 5,800,000; and thinks it extremely probable that in 1808 it exceeded 6,500,000.

The causes which retard the increase of numbers in Mexico are the small-pox; a disease called by the Indians matlazahuatl; and famine. The first of these, which was introduced in 1520, seems to exert its power at periods of 17 or 18 years. In 1763, and in 1779, it committed dreadful ravages, having carried off during the latter, in the capital alone, more than 9000 persons. In 1797 it was less destructive, chiefly in consequence of the zeal with which inoculation was propagated; between 50,000 and 60,000 individuals having undergone the operation. The vaccine method was introduced in various parts of Mexico and South America at the commencement of the present century. Humboldt mentions a curious circumstance, tending to show that the discovery of our celebrated countryman, Dr Jenner, had long been known to the country people among the Andes of Peru. A negro slave, who had been inoculated for the small-pox, showed no symptom of the disease, and when the practitioners were about to repeat the operation, told them he was certain that he should never take it; for, when milking cows in the mountains, he had been affected with cutaneous eruptions, caused, as the herdsmen said, by the contact of pustules sometimes found on the udders.

The frightful distemper called matlazahuatl, which is peculiar to the Indian race, seldom appears more than once in a century. It bears some resemblance to the yellow fever or black vomiting, which, however, very seldom attacks the natives. The extent of its ravages is not known with any degree of certainty, and it has not yet been submitted to medical investigation. Torquedama asserts that in 1545 it destroyed 800,000, and 2,000,000 in 1576; but these estimates are considered by Humboldt as greatly exaggerated.

A third obstacle to the progress of population in New Spain is famine. The American Indians, naturally indolent, contented with the smallest quantity of food on which life can be supported, and living in a fine climate, merely cultivate as much maize, potatoes, or wheat, as is necessary for their own maintenance, or at most for the additional consumption of the adjacent towns and mines. The inhabitants of Mexico have increased in a greater ratio than the means of subsistence, and accordingly, whenever the crops fall short of the demand, or are damaged by drought or other local causes, famine ensues. With want of food comes disease; and these visitations, which are of not unfrequent occurrence, are very destructive.

The working of the mines has also contributed to the depopulation of America. At the period of the conquest many Indians perished from excessive toil, and, as they were forced from their homes to distant places, they usually died without leaving progeny. In New Spain, however, such labour has been free for many years. The number employed in it does not exceed 28,000 or 30,000, and the mortality among them is not much greater than in other classes.

The Mexican population consists of the same elements as that of the other Spanish colonies. Seven races are distinguished:—1. *Gachupines,* or persons born in Europe; 2. Spanish *Creoles,* or Whites of European extraction born in America; 3. *Mestizoes,* descendants of Whites and Indians; 4. *Mulattoes,* descendants of Whites and Negroes; 5. *Zambos,* descendants of Negroes and Indians; 6. *Indians* of the indigenous race; and, 7. African *Negroes.*

The Indians appear to constitute at least two-fifths of the whole. Humboldt seems to favour the opinion, that the Aztecs, who inhabited New Spain at the period of the conquest, may have been of Asiatic origin. As the migrations of the American tribes have always taken place from north to south, the native population of this country must necessarily consist of very heterogeneous elements. The number of languages exceeds 20; and of these fourteen have tolerably complete grammars and dictionaries. Most of these tongues, so far from being only dialects of the same, as some authors have asserted, present as little affinity to each other as the Greek and the German. The variety spoken by the indigenous inhabitants of America forms a very striking contrast with the small number used in Asia and Europe. The Aztec or Mexican is the most widely distributed.

The Indians of New Spain bear a general resemblance to those of Florida, Canada, Peru, and Brazil. They have the same dingy copper colour, straight and smooth hair, deficient beard, squat body, elongated and oblique eyes, prominent cheekbones, and thick lips. But although the American tribes have thus a certain uniformity of character, they differ as much from each other as the numerous varieties of the European or Caucasian race. Those who live in this province have a more swarthy complexion than the inhabitants of the warmest parts of the South. They have also a much more abundant beard than the other tribes, and in the neighbourhood of the capital they even wear small moustaches. Pursuing a quiet and indolent life, and accustomed to uniform nourishment of a vegetable nature, they would no doubt attain a very great longevity were they not extremely addicted to drunkenness. They exist in a state of great moral degradation, being entirely destitute of religion, although they have exchanged their original rites for those of Catholicism. The men are grave, melancholic, and taciturn; forming a striking contrast to the negroes, who for this reason are preferred by the Indian women. Long habituated to slavery, they patiently suffer the privations to which they are frequently subjected; opposing to them only a degree of cunning, veiled under the appearance of apathy and stupidity. Although destitute of imagination, they are remarkable for the facility with which they acquire a knowledge of languages; and, notwithstanding their usual taciturnity, they become loquacious and eloquent when excited by important occurrences. It is unnecessary to speak of the negroes, of whom there are very few in Mexico, their character being the same as in other countries where slavery is permitted.

No city of the New Continent, not even excepting those of the United States, possesses more important scientific establishments than Mexico. Of these Humboldt mentions particularly the School of Mines, the Botanic Garden, which has however fallen into a state of neglect, and the Academy of Fine Arts. The influence of this institution is perceptible in the symmetry of the buildings which adorn the capital.

New Spain is divided into 15 districts, which he arranges as follows:—

I. In the TEMPERATE ZONE—82,000 square leagues; 677,000 inhabitants, or 8 to the square league—(1,059,193 square miles; inhabitants $\frac{8}{13}$ to the square mile).

A. Northern Region, in the interior.
1. Province of New Mexico, along the Rio del Norte, to the north of the parallel of 31°.

2. Intendancy of New Biscay, to the south-west of the Rio del Norte, on the central table-land.
B. North-western Region, in the vicinity of the Pacific Ocean.
   3. Province of New California, on the north-west coast of North America.
   4. Province of Old California, the southern extremity of which enters the torrid zone.
   5. Intendancy of La Sonora, which also passes the tropic.
C. North-eastern Region, adjoining the Gulf of Mexico.
   6. Intendancy of San Luis Potosi.

II. In the TORRID ZONE—36,500 square leagues; 5,160,000 inhabitants, or 141 to the square league—(471,470 square miles; inhabitants 11 to the square mile).

D. Central Region.
   7. Intendancy of Zacatecas.
   8. Intendancy of Guadalaxara.
   9. Intendancy of Guanaxuato.
   10. Intendancy of Valladolid.
   11. Intendancy of Mexico.
   12. Intendancy of Puebla.
   13. Intendancy of Vera Cruz.
E. South-western Region.
   14. Intendancy of Oaxaca.
   15. Intendancy of Merida.

Without attempting to present an analysis of our author's statistical account of these different provinces, we shall select from his descriptions those parts which may prove most interesting to the general reader.

1. The intendancy of Mexico is entirely within the torrid zone. More than two-thirds of it are mountainous, and contain extensive plains elevated from 2131 to 2451 feet above the sea. Only one summit, the Nevado de Tolucca, 15,158 feet in height, enters the region of perpetual snow.

The valley of Mexico, or Tenochtitlan, which is of an oval form, is situated in the centre of the cordillera of Anahuac, and is 63 miles in length by

43 in breadth. It is surrounded by a ridge of mountains, more elevated on the southern side, where it is confined by the great volcanoes of La Puebla, Popocatepetl, and Iztaccihuatl. The capital stands in the immediate vicinity of one of the great lakes which exist in this beautiful valley, although formerly it was placed on an island in that sheet of water, and communicated with the shore by three great dikes. This city is represented by Humboldt as one of the finest ever built by Europeans in either hemisphere, and all travellers agree in admiring its beauty. " From an eminence," says Captain Lyon in his interesting Journal, " we came suddenly in sight of the great valley of Mexico, with its beautiful city appearing in the centre surrounded by diverging shady paséos, bright fields, and picturesque haciendas. The great lake of Tezcuco lay immediately beyond it, shaded by a low floating cloud of exhalations from its surface, which hid from our view the bases of the volcanoes of Popocatepetl and Iztaccihuatl; while their snowy summits, brightly glowing beneath the direct rays of the sun, which but partially illumined the plains, gave a delightfully novel appearance to the whole scene before me. I was, however, at this distance, disappointed as to the size of Mexico; but its lively whiteness and freedom from smoke, the magnitude of the churches, and the extreme regularity of its structure, gave it an appearance which can never be seen in a European city, and declare it unique, perhaps unequalled in its kind."

The ground it occupies is every where perfectly level, the streets are regular and broad, the architecture generally of a very pure style, and many of the buildings are remarkably beautiful. Two kinds of hewn stone, a porous amygdaloid and a glassy felspar porphyry, are used. The houses are not loaded with decorations, nor disfigured by wooden balconies and galleries. The roofs are terraced; and the streets, which are clean and well lighted, have very broad pavements. The water of the lake is brackish, as is that of all the wells; but the city is supplied by two fine aqueducts. The objects which generally attract the notice of travellers are, 1. The cathedral, which has two towers ornamented with pilasters and statues; 2. The treasury; 3. The convents, of which the most distinguished is that of St Francis; 4. The hospital; 5. The acordada, a fine building, of which the prisons are spacious and well aired; 6. The school of mines; 7. The botanical garden; 8. The university; 9. The academy of fine arts; 10. The equestrian statue of Charles IV. in the great square.

Few remains of ancient monuments are to be found in the town or its vicinity. Of those that exist, the chief are the ruins of the Aztec dikes and aqueducts; the sacrificial stone, adorned with a relievo representing the triumph of a Mexican king; the great calendar in the plaza mayor; the colossal statue of the goddess Teoyaomiqui in one of the galleries of the university; the Aztec manuscripts or hieroglyphical pictures preserved in the house of the viceroys; and the foundations of the palace belonging to the sovereigns of Alcolhuacan at Tezcuco.

The only remarkable antiquities in the valley of Mexico are the remains of the two pyramids of San Juan de Teotihuacan, to the north-east of the lake of Tezcuco, consecrated to the sun and moon. One of these in its present state is a hundred and fifty feet in height, the other a hundred and forty-four. The interior is clay mixed with small stones, while the facings are of porous amygdaloid, and they are surrounded by a group of smaller elevation, disposed in a regular series. Another ancient object worthy of notice is the military entrenchment of Xochicalco, to the S.S.W. of the town of Cuernavaco, near Teteama. It consists of a hill 387 feet high, surrounded by ditches or trenches, and divided into five terraces covered with masonry; the whole forming a truncated pyramid, the four faces of which correspond to the four cardinal points. The porphyritic stones are adorned with hieroglyphical figures, among which are crocodiles, and men sitting cross-legged in the Asiatic manner. Other relics and places connected with the history of the conquest are shown to the stranger; but of these it is unnecessary to speak.

Our author estimates the population of Mexico as follows:—

	Inhabitants.
White Europeans,	2,500
White Creoles,	65,000
Copper-coloured natives,	33,000
Mestizoes, mixture of Whites and Indians,	26,500
Mulattoes,	10,000
	137,000

The annual number of births for a mean term of 100 years is 5930, and that of deaths 5050; while in New Spain in general, the relation of the births to the population is as 1 to 17, and that of the deaths as 1 to 30, so that the mortality in the capital appears much greater. The great conflux of sick persons to the hospitals, and on the other hand the celibacy of the numerous clergy, the progress of luxury, and other causes, induce this disproportion.

According to researches made by the Count de Revillagigedo, the consumption of Mexico in 1791 was as follows:—

I. ANIMAL FOOD.

Oxen,	16,300	Fowls,	1,255,340
Calves,	450	Ducks,	125,000
Sheep,	278,923	Turkeys,	205,000
Hogs,	50,676	Pigeons,	65,300
Kids and Rabbits,	24,000	Partridges,	140,000

II. GRAIN.

Maize, or Indian corn—cargas of 3 fanegas, 117,224 = 545,219 I. S. bushels.
Barley—cargas, 40,219 = 187,062 I. S. bushels.
Wheat flour, cargas of 12 arrobas, 130,000 = 353,229 cwt.

III. LIQUIDS.

Pulque, the fermented juice of agave—cargas, 294,790 = 800,987 cwts.
Wine and vinegar barrels of $4\frac{1}{2}$ arrobas, 4,507 = 71,756 I. S. galls.
Brandy—barrels, 12,000 = 191,052 I. S. galls.
Spanish oil—arrobas of 25 pounds, 5,585 = 15,530 I. S. galls.

The market is abundantly supplied with vegetables of numerous kinds, which are brought in every morning by the Indians in boats. Most of these are cultivated on the chinampas or gardens, some of which float upon the neighbouring sheet of water, while others are fixed in the marshy grounds.*

The surface of the four principal lakes in the valley of Mexico occupies nearly a tenth of its extent;

---

* " These are long narrow stripes of ground redeemed from the surrounding swamp, and intersected by small canals. They all appeared to abound in very fine vegetables, and lively-foliaged poplars generally shadowed their extremities. The little gardens constructed on bushes or wooden rafts no longer exist in the immediate vicinity of Mexico; but I learnt that some may yet be seen at Inchimilco, a place near San Augustin de las Cuevas."—*Captain Lyon's Journal of a Residence and Tour in the Republic of Mexico*, vol. ii. p. 110.

or 168 square miles. The lake of Xochimilco contains $49\frac{1}{2}$, that of Tezcuco 77, of San Christobal $27\frac{1}{2}$, and of Zumpango $9\frac{9}{10}$, square miles. The valley itself is a basin enclosed by a wall of porphyritic mountains, and all the water furnished by the surrounding cordilleras is collected in it. No stream issues from it excepting the brook of Tequisquiac, which joins the Rio de Tula. The lakes rise by stages in proportion to their distance from its centre, or, in other words, from the site of the capital. Next to the lake of Tezcuco, Mexico is the least elevated point of the valley, the plaza mayor or great square being only 1 foot 1 inch higher than the mean level of its water, which is $11\frac{3}{4}$ feet lower than that of San Christobal. Zumpango, which is the most northern, is 29·211 inches higher than the surface of Tezcuco; while that of Chalco, at the southern extremity, is only 3·632 feet more elevated than the great square of Mexico.

In consequence of this peculiarity the city has for a long series of ages been exposed to inundations. The lake of Zumpango, swelled by an unusual rise of the Rio de Guautitlan, flows over into that of San Christobal, which again bursts the dike that separates it from Tezcuco. The water of this last is consequently augmented, and flows with impetuosity into the streets of Mexico. Since the arrival of the Spaniards the town has experienced five great floods, the latest of which happened in 1629. In more recent periods there have been several alarming appearances, but the city was preserved from any actual loss by the desague or canal, which was formed for the purpose.

The situation of the capital is more exposed to danger, because the bed of the lake is progressively rising in consequence of the mud carried into it, and the difference between it and the level of the plain diminishing. Previous to the conquest, and for some time after, it was defended by dikes; but this method having been found ineffectual, the viceroy in 1607 employed Enrico Martinez, a native of Germany, to effect the evacuation of the lakes. After making an exact survey of the valley he presented two plans for canals, the one to empty those of Tezcuco, Zumpango, and San Christobal, the other to drain that of Zumpango alone. The latter scheme was adopted, and in consequence, the famous subterraneous gallery of Nochistongo was commenced on the 28th November 1607. Fifteen thousand Indians were employed, and after eleven months of continued labour the work was completed. Its length was more than 21,654 feet, its breadth 11·482, and its height 13·780. On the opposite side of the hill of Nochistongo is the Rio de Tula, which runs into the Rio de Panuco, and from the northern or further extremity of the gallery an open trench, 28,216 feet long, was cut to carry the water to the former river. Soon after the current began to flow through this artificial channel, it gradually occasioned depositions and erosions, so that it became necessary to support the roof, which was composed of marl and clay. For this purpose wood was at first employed, and afterwards masonry; but the arches being soon undermined, the passage at length was obstructed.

Several plans were now proposed, and in 1614 the court of Madrid sent to Mexico a Dutch engineer, Adrian Boot, who advised the construction of great dikes after the Indian plan. A new viceroy, however, having recently arrived, who had never witnessed the effects of an inundation, ordered Martinez to stop up the subterraneous passage, and make the water of the upper lakes return to the bed of the Tezcuco, that he might see if the danger were really so great as it had been represented. Being convinced that it was so, he ordered the German to recommence his operations in the gallery. The engineer accordingly proceeded to clear it, and continued working until the 20th June 1629, when finding the mass of water too great to be received by this narrow outlet, he closed it in order to prevent its destruction. In the morning the city of Mexico was flooded to the depth of three feet, and, contrary to expectation, remained in that state for five years. In this interval various plans were proposed for draining the neighbouring lake, although none of them was carried into effect; but the inundation at length subsided in consequence of a succession of earthquakes.

Martinez, who had been imprisoned from a belief that he had closed the gallery for the purpose of affording the incredulous a proof of the utility of his work, was now set at liberty, and constructed the dike of San Christobal. He was ordered to enlarge the gallery; but the operations were conducted with very little energy, and in the end it was determined to abandon the plan, to remove the top of the vault, and to convert it into an open passage by cutting through the hill. A lawyer, named Martin de Solis, undertook the management of this enterprise; though it required nearly two centuries to complete the work; the canal not being opened in its whole length until 1789. As it now appears, it is stated by Humboldt to be one of the most gigantic hydraulic operations executed by man. Its length is 67,537 feet, its greatest depth 197, and its greatest breadth 361.

The safety of the capital depends, 1st, On the stone dikes which prevent the water of the lake of Zumpango from passing into that of San Christobal, and the latter from flowing into the Tezcuco; 2d, On the dikes and sluices which prevent the lakes of Chalco and Xochimilco from overflowing; 3d, On the great cut of Enrico Martinez, by which the Rio de Guautitlan passes across the hills into the valley of Tula; and, 4th, On the canals by which the Zumpango and San Christobal may be completely drained. These means however, expensive and numerous as they must appear, are insufficient to secure it against inundations proceeding from the north and north-west; and our author asserts, that it will continue exposed to great risks until a canal shall be directly opened from the lake of Tezcuco.

The intendancy of Mexico contains, besides the capital, several towns of considerable size, of which the more important are, Tezcuco, Acapulco, Tolucca, and Queretaro, the latter having a population of thirty-five thousand.

2. The government of Puebla is wholly situated in the torrid zone, and is bounded on the north-east by that of Vera Cruz, on the south by the ocean, on the east by the province of Oaxaca, and on the west by that of Mexico. It is traversed by the cordilleras of Anahuac, and contains the highest mountain in New Spain, the volcano of Popo-

catepetl. A great portion, however, consists of an elevated plain, on which are cultivated wheat, maize, agave, and fruit-trees.

The population is concentrated on this table-land, extending from the eastern slope of the Nevados, or Snowy Mountains, to the vicinity of Perote. It exhibits remarkable vestiges of ancient Mexican civilisation. The great pyramid of Cholula has a much larger base than any edifice of the kind in the Old Continent, its horizontal breadth being not less than 1440 feet; but its present height is only fifty-nine yards, while the platform on its summit has a surface of 45,210 feet.

At the village of Atlixco is seen a cypress (*Cupressus disticha*) 76 feet in circumference, which is probably one of the oldest vegetable monuments on the globe.* There are very considerable salt-works in this intendancy, and a beautiful marble is quarried in the vicinity of Puebla. The principal towns are that just named, containing a population of 67,800, Cholula, Tlascala, and Atlixco.

3. The intendancy of Guanaxuato, situated on the ridge of the cordillera of Anahuac, is the most populous in New Spain, and contains three cities, Guanaxuato, Celayo, and Salvatierra, four towns, 37 villages, and 448 farms or haciendas. It is in

---

* "On entering the gardens of Chapultepec (near Mexico), the first object that strikes the eye is the magnificent cypress (*Sabino Ahuahuete*, or *Cupressus disticha*), called the Cypress of Montezuma. It had attained its full growth when that monarch was on the throne (1520), so that it must now be at least 400 years old, yet it still retains all the vigour of youthful vegetation. The trunk is 41 feet in circumference, yet the height is so majestic as to make even this enormous mass appear slender."—*Ward's Mexico in* 1827, vol. ii. p. 230. The same author mentions another cypress, 38 feet in girth, and of equal height to that of Montezuma.

general highly cultivated, and possesses the most important mines in that section of the New World.

4. The intendancy of Valladolid is bounded on the north by the Rio de Lerma; on the east and northeast by that of Mexico; on the south by the district of Guanaxuato; and on the west by the province of Guadalaxara. Being situated on the western declivity of the cordillera of Anahuac and intersected by hills and beautiful valleys, it in general enjoys a mild and temperate climate. The volcano of Jorullo, already described, is situated in this intendancy, which has three cities, three towns, and 263 villages. The southern part is inhabited by Indians.

5. The province of Guadalaxara is bounded on the north by the governments of Sonora and Durango, on the east by those of Zacatecas and Guanaxuato, on the south by the district of Valladolid, and on the west by the Pacific Ocean. Its greatest breadth is 345 miles, and its greatest length 407. It is crossed from east to west by the Rio de Santiago, which is of considerable size. The eastern portion consists of the elevated platform and western declivity of the cordilleras of Anahuac. The maritime parts are covered with forests which abound in excellent timber. The volcano of Colima, situated in this district, is the most western of those of New Spain. It frequently throws up ashes and smoke; but its height is not so great as to carry its summit into the region of perpetual snow. The most remarkable towns are, Guadalaxara, which has a population of 19,500, San Blas, a port at the mouth of the Santiago, and Compostella.

6. The intendancy of Zacatecas, bounded on the north by Durango, on the east by San Luis Potosi,

on the south by Guanaxuato, and on the west by Guadalaxara, is 293 miles in length, and 176 in breadth. The table-land, which forms its central part, is composed of syenite and primitive slate. Near Zacatecas are nine small lakes abounding in muriate and carbonate of soda. This district is very thinly peopled, although the town has 33,000 inhabitants.

7. The intendancy of Oaxaca is one of the most delightful countries in the New Continent, possessing great fertility of soil and salubrity of climate. It is bounded on the north by Guatimala; on the west by the province of Puebla; and on the south by the Pacific Ocean. The mountainous parts are composed of granite and gneiss. The vegetation is every where exceedingly beautiful. At the village of Santa Maria del Tule, ten miles east of the capital, there is an enormous trunk of *Cupressus disticha*, 118 feet in circumference, though it seems rather to be formed of three stems grown into one.

The most remarkable object in this district is the palace of Mitla, the walls of which are decorated with grecques and labyrinths in mosaic, resembling the ornaments of Tuscan vases. It consists of three edifices, and is moreover distinguished from other ancient Mexican buildings by six porphyritic columns which support the ceiling of a vast hall. These pillars have neither base nor capital; each exhibits a single block of stone, and the height is about sixteen feet. Oaxaca, the principal town, contained, in the year 1792, twenty-four thousand inhabitants. Some of the mines are very productive.

8. The intendancy of Merida comprehends the great peninsula of Yucatan, situated between the

Bay of Campeachy and that of Honduras. It is bounded on the south by Guatimala, on the east by the province of Vera Cruz, and on the west by the English establishments which extend from the mouth of the Rio Hondo to the north of the Bay of Hanover. This peninsula is a vast plain, intersected by a chain of hills; and though one of the warmest, it is at the same time one of the healthiest provinces of equinoctial America. The latter circumstance is to be attributed to the extreme dryness of the soil and atmosphere. No European grain is produced; but maize, jatropha, and dioscorea are cultivated in abundance. The Hœmatoxylon or Campeachy wood abounds in several districts. Merida, the capital, has a population of 10,000.

9. The government of Vera Cruz extends along the Mexican Gulf from the Rio Baraderas to the great river of Panuco. The western part forms the declivity of the cordilleras of Anahuac, from whence, amid the regions of perpetual snow, the inhabitants descend in a day to the burning plains of the coast. In this district are displayed in a remarkable manner the gradations of vegetation, from the level of the sea to those elevated summits which are visited with perennial frost. In ascending, the traveller sees the physiognomy of the country, the aspect of the sky, the form of the plants, the figures of animals, the manners of the inhabitants and the kind of cultivation followed by them, assuming a different appearance at every step. Leaving the lower districts, covered with a beautiful and luxuriant vegetation, he first enters that in which the oak appears, where he has no longer cause to dread the yellow fever, so fatal on the coasts. Forests of liquidambar, near Xalapa, an-

nounce by their freshness the elevation at which the strata of clouds, suspended over the ocean, come in contact with the basaltic summits of the cordilleras. A little higher the banana ceases to yield fruit. At the height of San Miguel pines begin to mingle with the oaks, which continue as far as the plains of Perote, where the cereal vegetation of Europe is seen. Beyond this, the former alone cover the rocks, the tops of which enter the region of perpetual frigidity.

At the foot of the cordillera, in the evergreen forests of Papautla, Nautla, and S. Andre Tuxtla, grows the vanilla, the fruit of which is used for perfuming chocolate. The beautiful convolvulus, whose root furnishes the jalap of the apothecaries, grows near the Indian villages of Colipa and Misautla. The pimento-myrtle is produced in the woods which extend towards the river of Baraderas. On the declivities of Orizaba, tobacco of excellent quality is cultivated; and the sarsaparilla grows in the moist and shady ravines. Cotton and sugar of excellent quality are produced along the greater part of the coast.

In this intendancy are two colossal summits,— the volcano of Orizaba, which after Popocatepetl is the highest in New Spain, and the Cofre de Perote, which is nearly 1312 feet more elevated than the Peak of Teneriffe. In its northern part, near the Indian village of Papautla, is a pyramidal edifice of great antiquity situated in the midst of a thick forest. It is not constructed of bricks, or clay mixed with stone, and faced with amygdaloid, like those of Cholula and Tectihuacan; on the contrary, the materials employed have been immense blocks of porphyry. The base is an exact square, 82 feet on each side, and the perpendicular height seems to be about sixty. It is composed of several stages, of which some are still distinguishable. A great stair of 57 steps conducts to the truncated summit.

The most remarkable cities are Vera Cruz, Perote, Cordoba, and Orizaba. The first of these, the centre of European and West Indian commerce, is beautifully and regularly built; but it is situated in an arid plain destitute of running water and partly covered with shifting sand-hills, which contribute to increase the suffocating heat of the air. In the midst of these downs are marshy lands covered with rhizophoræ and other plants. No stones for architectural purposes are to be found near the city, which is entirely constructed of coral rock drawn from the bottom of the sea. The water is very bad, and is obtained either by digging in the sandy soil, or by collecting the rain in cisterns.

Xalapa, the population of which is estimated at 13,000, occupies a very romantic situation at the foot of the basaltic mountain of Macultepec, surrounded by forests of styrax, piper, melastomæ, and tree-ferns. The sky is beautiful and serene in summer, but from December to February it has a most melancholy aspect, and, whenever the north wind blows, is overcast to such a degree that the sun and stars are frequently invisible for two or three weeks together. Some of the merchants of Vera Cruz have country-houses at Xalapa, where they enjoy a cool and agreeable retreat, while the coast is almost uninhabitable, on account of the intense heats, the mosquitoes, and the yellow fever.

10. The captaincy of San Luis Potosi embraces

the whole north-eastern part of New Spain, and is extremely diversified in its character. The only portion which is cold and mountainous is that adjoining the province of Zacatecas, and in which are the rich mines of Charcas, Guadalcagar, and Catorce. There is a great extent of low ground, partly cultivated, but for the most part barren and uninhabited. Its coast line is more than 794 miles in length; but hardly any commerce enlivens it, owing to the deficiency of harbours. The mouths of the rivers, too, are blocked up by bars, necks of land, and long islands running parallel to the coast.

11. New Biscay or Durango occupies a greater space of ground than Great Britain and Ireland, though its population does not exceed 160,000. It is bounded on the south by Zacatecas and Guadalaxara; on the south-east by San Luis; and on the west by Sonora. On the northern and eastern sides, for more than 690 miles, it borders on an uncultivated country inhabited by independent Indians. This intendancy comprehends the northern extremity of the great table-land of Anahuac, which declines towards the Rio Grande del Norte.

12. The province of Sonora is still more thinly peopled than Durango. It extends on the shores of the Gulf of California more than 966 miles.

13. New Mexico, which is very sparingly inhabited, stretches along the Rio Norte, and has a remarkably cold climate.

14. Old California equals England in extent of territory, but has only a population of 9000. The soil of this peninsula is parched and sandy, and the vegetation feeble; but the sky is constantly clear and of a deep blue; the light clouds which sometimes ap-

pear presenting at sunset the most beautiful shades of violet, purple, and green. A chain of mountains, the highest of which is about 5000 feet, runs through the centre of the peninsula, and is inhabited by animals resembling the mouflon of Sardinia, which the Spaniards call wild sheep. The principal attraction which California has afforded to Europeans since the 16th century is the great quantity of pearls found in it, and which, although frequently of an irregular form, are large and of a very beautiful water. At the present day, however, this fishery is almost entirely abandoned.

15. New California is a long and narrow country, identifying itself with the shore of the Pacific Ocean from the isthmus of Old California to Cape Mendocino. It is extremely picturesque, and enjoys a fertile well-watered soil, with a temperate climate. Wheat, barley, maize, beans, and other useful plants, thrive well, as do the vine and olive; but the population is scanty compared to the territory. A cordillera of small elevation runs along the coast, and the forests and prairies are filled with deer of gigantic size.

CHAPTER XXV.

Statistical Account of New Spain continued.

Agriculture of Mexico—Banana, Manioc, and Maize—Cereal Plants—Nutritive Roots and Vegetables—Agave Americana—Colonial Commodities—Cattle and Animal Productions.

A COUNTRY extending from the sixteenth to the thirty-seventh degree of latitude, and presenting a great variety of surface, necessarily affords numerous modifications of climate. Such is the admirable distribution of heat on the globe, that the strata of the atmosphere become colder as we ascend, while those of the sea are warmest near the surface. Hence, under the tropics, on the declivities of the cordilleras, and in the depths of the ocean, the plants and marine animals of the polar regions find a temperature suited to their development. It may easily be conceived that, in a mountainous country like Mexico, having so great a diversity of elevation, temperature, and soil, the variety of indigenous productions must be immense; and that most of the plants cultivated in other parts of the globe may there find situations adapted to their nature.

There, however, the principal objects of agriculture are not the productions which European luxury draws from the West India Islands, but the grasses, nutritive roots, and the agave. The appearance of the land proclaims to the traveller that the natives are nourished by the soil, and that they are independent of foreign commerce. Yet agriculture is by no means so flourishing as might be expected from its natural resources, although considerable improvement has been effected of late years. The depressed state of cultivation, it is true, has been attributed to the existence of numerous rich mines; but Humboldt, on the contrary, maintains that the working of these ores has been beneficial in causing many places to be improved which would otherwise have remained steril. When a vein is opened on the barren ridge of the cordilleras, the new colonists can only draw the means of subsistence from a great distance. Want soon excites to industry, and farms begin to be established in the neighbourhood. The high price of provisions indemnifies the cultivator for the hard life to which he is exposed, and the ravines and valleys become gradually covered with food. When the mineral treasures are exhausted, the workmen no doubt emigrate, so that the population is diminished; but the settlers are retained by their attachment to the spot in which they have passed their childhood. The Indians, moreover, prefer living in the solitudes of the mountains remote from the whites, and this circumstance tends to increase the number of inhabitants in such districts.

In describing the vegetable productions of New Spain, our author begins with those which form the principal support of the people, then treats of the class which affords materials for manufacture, and ends with such as constitute objects of commerce.

The banana (*Musa paradisiaca*) is to the inhabitants of the torrid zone what the cereal grasses, —wheat, barley, and rye,—are to Western Asia and

Europe, and what the numerous varieties of rice are to the natives of India and China. Forster and other naturalists have maintained, that it did not exist in America previous to the arrival of the Spaniards, but that it was imported from the Canary Islands in the beginning of the 16th century; and in support of this opinion may be adduced the silence of Columbus, Alonzo Negro, Pinzon, Vespucci, and Cortes, with respect to it. This circumstance, however, only proves the inattention of these travellers to the productions of the soil; and it is probable that the Musa presented several species indigenous to different parts of both continents. The space favourable to the cultivation of this valuable plant in Mexico is more than 50,000 square leagues, and has nearly a million and a half of inhabitants. In the warm and humid valleys of Vera Cruz, at the foot of the cordillera of Orizaba, the fruit occasionally exceeds 11·8 inches in circumference, with a length of seven or eight. A bunch sometimes contains from 160 to 180, and weighs from 66 to 88 ℔. avoirdupois.

Humboldt doubts whether there is any other plant on the globe which, in so small a space of ground, can produce so great a mass of nutriment. Eight or nine months after the sucker has been inserted in the earth the banana begins to form its clusters, and the fruit may be gathered in less than a year. When the stalk is cut, there is always found among the numerous shoots which have put forth roots one that bears three months later. A plantation is perpetuated without any other care than that of cutting the stems on which the fruit has ripened, and giving the earth a slight dressing. A spot of 1076 feet may contain at least from thirty to forty plants, which, in the space of a year, at a very moderate calculation, will yield more than 4410 ℔. avoirdupois of nutritive substance. Our author estimates, that the produce of the banana is to that of wheat as 133 : 1, and to that of potatoes as 44 : 1.

In America numerous preparations are made of this fruit, both before and after its maturity. When fully ripe it is exposed to the sun, and preserved like our figs; the skin becoming black, and exhaling a peculiar odour like that of smoked ham. This dry banana (*platano passado*), which is an object of commerce in the province of Mechoacan, has an agreeable taste, and is a very wholesome article of food. Meal or flour is obtained from it, by being cut into slices, dried in the sun, and pounded.

It is calculated that the same extent of ground in Mexico on which the banana is raised is capable of maintaining fifty individuals, whereas in Europe, under wheat it would not furnish subsistence for two; and nothing strikes a traveller more than the diminutive appearance of the spots under culture round a hut which contains a numerous family.

The region where it is cultivated produces also the valuable plant (*Jatropha*), of which the root, as is well known, affords the flour of manioc, usually converted into bread, and furnishes what the Spanish colonists call *pan de tierra caliente*. This vegetable is only successfully grown within the tropics, and in the mountainous region of Mexico is never seen above the elevation of 2625 feet. Two kinds are raised, the sweet and the bitter. The root of the former may be eaten without danger, while that of the latter is a very active poison. Both may be made into bread; but the bitter is preferred for this purpose, the poisonous juice being carefully separated from the fecula, called cassava, before making the dough. Raynal asserted that the manioc was transported from Africa to America to serve for the maintenance of the negroes; but our author shows that it was cultivated there long before the arrival of Europeans on that side of the Atlantic. The bread made of it is very nutritive; but, being extremely brittle, it does not answer for distant carriage. The fecula, however, grated, dried, and smoked, is used on journeys. The root loses its poisonous qualities on being boiled, and in this state the decoction is used as a sauce, although serious accidents sometimes happen when it has not been long enough exposed to heat. The husbandry of it, we may observe, requires more care than that of the banana. In this respect it resembles the potato; and the roots are ripe in seven or eight months after the slips have been planted.

The same region produces maize, the cultivation of which is more extensive than that of the banana and manioc. Advancing towards the central plains, we meet with fields of this important plant all the way from the coast to the valley of Tolucca, which is upwards of 9186 feet above the sea. Although a great quantity of other grain is produced in Mexico, this must be considered as the principal food of the people, as well as of most of the domestic animals, and the year in which the maize-harvest fails is one of famine and misery to the inhabitants. There is no longer a doubt among botanists that this plant is of American origin, and that the Old Continent received it from the New.

It does not thrive in Europe where the mean

temperature is less than 44° or 46°; and on the cordilleras of New Spain rye and barley are seen to vegetate vigorously where the cultivation of maize would not be attended with success. On the other hand, the latter thrives in the lowest plains of the torrid zone, where wheat, barley, and rye, are not found. Hence we cannot be surprised to hear that it occupies a much greater extent in equinoctial America than the grains of the Old Continent.

The fecundity of the Mexican variety is astonishing. Fertile lands usually afford a return of 300 or 400 fold, and in the neighbourhood of Valladolid a harvest is considered defective when it yields only 130 or 150. Even where the soil is most steril the produce varies from sixty to eighty. The general estimate for the equinoctial region of Mexico may be considered as a hundred and fifty.

Of all the gramina cultivated by man, none is so unequal as this in its produce, as it varies in the same field according to the season from forty to 200 or 300 for one. If the harvests are good, the agriculturist makes his fortune more rapidly than with any other grain; but frightful dearths sometimes occur, when the natives are obliged to feed on unripe fruit, cactus-berries, and roots. Diseases arise in consequence; and these famines are usually attended with a great mortality among the children. Fowls, turkeys, and even cattle suffer, so that the traveller can find neither eggs nor poultry. Scarcities of less severity are not uncommon, and are especially felt in the mining districts, where the vast numbers of mules employed in the process of amalgamation annually consume an enormous quantity of maize.

Numerous varieties of food are derived from this

plant. The ear is eaten raw or boiled. The grain when beaten affords a nutritive bread called arepa, and the meal is employed in making soups or gruels, which are mixed with sugar, honey, and sometimes even pounded potatoes. Many kinds of drink are also prepared from it, some resembling beer, others cider. In the valley of Tolucca the stalks are squeezed between cylinders, and from the fermented juice a spiritous liquor, called *pulque de mahis*, is procured.

In favourable years Mexico yields a much larger quantity than is necessary for its own consumption; but as this grain affords less nutritive substance in proportion to its bulk than the corn of Europe, and as the roads are generally difficult, obstacles are presented to its transportation, which, however, will diminish when the country is more improved.

We come now to the cereal plants which have been conveyed from the Old to the New Continent. A negro slave of Cortes found among the rice, which served to maintain the Spanish army, three or four particles of wheat, which were sown, we may suppose, before the year 1500. A Spanish lady, Maria d'Escobar, carried a few grains to Lima, and their produce was distributed for three years among the new colonists, each receiving twenty or thirty seeds. At Quito the first European corn was sown near the convent of St Francis by Father Jose Rixi, a native of Flanders, and the monks still show, as a precious relic, the earthen vessel in which the original wheat came from Europe. " Why," asks our author, " have not men preserved every where the names of those who, in place of ravaging the earth, have enriched it with plants useful to the human race?"

The temperate region appears most favourable to the cultivation of the cerealia, or nutritive grasses known to the ancients, namely, wheat, spelt, barley, oats, and rye. In the equinoctial part of Mexico they are nowhere grown in plains of which the elevation is under 2625 feet; and on the declivity of the cordilleras between Vera Cruz and Acapulco they commence at the height of 3937. At Xalapa wheat is raised solely for the straw; for there it never produces seed, although in Guatimala grain ripens at smaller elevations.

Were the soil of New Spain watered by more frequent showers, it would be one of the most fertile portions of the globe. In the equinoctial districts of that country there are only two seasons,—the wet, from June or July to September or October, and the dry, which lasts eight months. The rains, accompanied with electrical explosions, commence on the eastern coast, and proceed westward, so that they begin fifteen or twenty days sooner at Vera Cruz than on the central plains. Sometimes they are seen, mixed with sleet and snow, in the elevated parts during November, December, and January, but they last only a few days. It is seldom that the inhabitants have to complain of humidity, and the excessive drought which prevails from June to September compels them in many parts to have recourse to artificial irrigation. In places not watered in this manner, the soil yields pasturage only till March or April, after which the south wind destroys the grass. This change is more felt when the preceding year has been unusually dry, and the wheat suffers greatly in May. The rains of June, however, revive the vegetation, and the fields immediately resume their verdure.

In lands carefully cultivated the produce is surprising, especially in those which are watered. In the most fertile part of the table-land between Queretaro and Leon, the wheat-harvest is 35 and 40 for 1; and several farms can even reckon on 50 or 60 for 1. At Cholulo the common return is from 30 to 40, but it frequently exceeds from 70 to 80 for 1. In the valley of Mexico maize yields 200, and wheat 18 or 20. The mean produce of the whole country may be stated at 20 or 25 for 1. M. Abad, a canon of the metropolitan church of Valladolid de Mechoacan, took at random from a field of wheat forty plants, when he found that each seed had produced forty, sixty, and even seventy stalks. The number of grains which the ears contained frequently exceeded 100 or 120, and the average amount appeared to be 90. Some even exhibited 160. A few of the elevated tracts, however, are covered with a kind of clay impenetrable by the roots of herbaceous plants, and others are arid and naked, in which the cactus and other prickly shrubs alone vegetate.

The following table exhibits the mean produce of the cereal plants in different countries of both continents:—

> In France, from 5 to six grains for 1.
> In Hungary, Croatia, and Sclavonia, from 8 to 10 grains.
> In La Plata, 12 grains.
> In the northern part of Mexico, 17 grains.
> In equinoctial Mexico, 24 grains.
> In the province of Pasto in Santa Fe, 25 grains.
> In the plain of Caxamarca in Peru, 18 to 20 grains.

The Mexican wheat is of the very best quality, and equals the finest Andalusian. At Havannah it enters into competition with that of the United States, which is considered inferior to it; and when greater facilities are afforded for exportation it will become of the highest importance to Europe. In Mexico grain can hardly be preserved longer than two or three years; but the causes of this decay have not been sufficiently investigated.

Rye and barley, which resist cold better than wheat, are cultivated on the highest regions, but only to a small extent. Oats do not answer well in New Spain, and are very seldom seen even in the mother-country, where the horses are fed on barley.

The potato appears to have been introduced into Mexico nearly at the same period as the cereal grasses of the Old Continent. It is certain that it was not known there before the arrival of the Spaniards, at which epoch it was in use in Chili, Peru, Quito, and New Grenada. It is supposed by botanists, that it grows spontaneously in the mountainous regions; but our author asserts that this opinion is erroneous, and that the plant in question is nowhere to be found uncultivated in any part of the cordilleras within the tropics. According to Molina it is a native of all the fields of Chili, where another species, the *Solanum cari*, still unknown in Europe, and even in Quito and Mexico, is grown; and M. Humboldt seems to consider that country as the original source of it. It is stated that Sir Walter Raleigh found it in Virginia in 1584; and a question arises, whether it arrived there from the north, or from Chili, or some other of the Spanish colonies. Our traveller seems to consider it not improbable that it had been conveyed from some of the Spanish colonies by the English themselves.

The plants cultivated in the highest and coldest parts of the Andes and Mexican cordilleras are potatoes, the *Tropæolum esculentum*, and the *Chenopodium quinoa*. The first of these are an important object in the latter country, as they do not require much humidity. The Mexicans and Peruvians preserve them for a series of years, by destroying their power of germinating by exposure to frost, and afterwards drying them,—a practice which our author thinks might be followed with advantage in Europe. He also recommends obtaining the seeds of the potatoes cultivated at Quito and Santa Fe, which are a foot in diameter, and superior in quality to those in the Old Continent. It is unnecessary to expatiate on the advantages derived from this invaluable root, the use of which now extends from the extremity of Africa to Lapland, and from the southern regions of America to Labrador.

The New World is very rich in plants with nutritive roots. Next to the manioc and the potato, the most important are the *oca*, the *batate*, and the *igname*. The first of these (*Oxalis tuberosa*) grows in the cold and temperate parts of the cordilleras. The igname (*Dioscorea alata*) appears proper to all the equinoctial regions of the globe. Of the batate (*Convolvulus batatas*), several varieties are raised. The *cacomite*, a species of Tigridia, the root of which yields a nutritive farina; numerous varieties of love-apples (*Solanum lycopersicum*); the earth pistachio or mani (*Arachis hypogœa*); and different species of pimento (*Capsicum*), are the other useful plants cultivated there.

The Mexicans now have all the culinary vegetables and fruit-trees of Europe; but it has become difficult to determine which of the former they possessed before the arrival of the Spaniards. It is certain, however, that they had onions, haricots, gourds, and several varieties of Cicer; and in general, if we consider the garden-stuffs of the Aztecs and the great number of farinaceous roots cultivated in Mexico and Peru, we shall see that they were not so poor in alimentary plants as some maintain.

The central table-land of New Spain produces the ordinary fruits of Europe in the greatest abundance; and the traveller is surprised to see the tables of the wealthy inhabitants loaded with the vegetable productions of both continents in the most perfect state. Before the invasion of the Spaniards, Mexico and the Andes presented several fruits having a great resemblance to those of Europe. The mountainous part of South America has a cherry, a nut, an apple, a mulberry, a strawberry, a rasp, and a gooseberry, which are peculiar to it. Oranges and citrons, which are now cultivated there, appear to have been introduced, although a small wild orange occurs in Cuba and on the coast of Terra Firma. The olive-tree answers perfectly in New Spain, but exists only in very small numbers.

Most civilized nations procure their drinks from the plants which constitute their principal nourishment, and of which the roots or seeds contain saccharine and amylaceous matter. There are few tribes, indeed, which cultivate these solely for the purpose of preparing beverages from them; but in the New Continent we find a people, who not only extracted liquors from the maize, the manioc, and bananas, but who raise a shrub of the family of the ananas for the express purpose of converting its juice into a spiritous liquor. This plant, the maguey (*Agave Americana*), is extensively reared as far as the Aztec language extends. The finest plantations of it seen by our traveller were in the valley of Tolucca and on the plains of Cholula. It yields the saccharine juice at the period of inflorescence only, the approach of which is anxiously observed. Near the latter place, and between Tolucca and

Cacanumacan, a maguey eight years old gives signs of developing its flowers. The bundle of central leaves is now cut, the wound is gradually enlarged and covered with the foliage, which is drawn close and tied at the top. In this wound the vessels seem to deposite the juice that would naturally have gone to expand the blossoms. It continues to run two or three months, and the Indians draw from it three or four times a-day. A very vigorous plant occasionally yields the quantity of 454 cubic inches a-day for four or five months. This is so much the more astonishing, that the plantations are usually in the most arid and steril ground. In a good soil the agave is ready for being cut at the age of five years; but in poor land the harvest cannot be expected in less than eighteen.

This juice or *honey* has an agreeable acid taste, and easily ferments on account of the sugar and mucilage which abound in it. This process, which is accelerated by adding a little old pulque, ends in three or four days; and the result is a liquor resembling cider, but with a very unpleasant smell like that of putrid meat. Europeans who can reconcile themselves to the scent prefer the *pulque* to every other liquor, and it is considered as stomachic, invigorating, and nutritive. A very intoxicating brandy, called mexical, is also obtained from it, and in some districts is manufactured to a great extent.

The leaves of the agave also supply the place of hemp and the papyrus of the Egyptians. The paper on which the ancient Mexicans painted their hieroglyphical figures was made of their fibres, macerated and disposed in layers. The prickles which terminate them formerly served as pins and nails to the Indians, and the priests pierced their arms and breasts with them in their acts of expiation.

The vine is cultivated in Mexico, but in so small a quantity that wine can hardly be considered as a product of that country; but the mountainous parts of New Spain, Guatimala, New Grenada, and Caraccas, are so well adapted for its growth, that at some future period they will probably supply the whole of North America.

Of colonial commodities, or productions which furnish raw materials for the commerce and manufacturing industry of Europe, New Spain affords most of those procured from the West Indies. The cultivation of the sugar-cane has of late years been carried to such an extent, that the exportation of sugar from Vera Cruz amounts to more than half a million of arrobas, or 12,680,000 lb. avoird.; which, at 3 piasters the arroba, are equal to 5,925,000 francs, or £246,875 sterling. It was conveyed by the Spaniards from the Canary Islands into St Domingo, from whence it was subsequently carried into Cuba and the province just named. Although the mean temperature best suited to it is 75° or 77°, it may yet be successfully reared in places of which the annual warmth does not exceed 66° or 68°; and as on great table-lands the heat is increased by the reverberation of the earth, it is cultivated in Mexico to the height of 4921 feet, and in favourable exposures thrives even at an elevation of 6562. The greatest part of the sugar produced in New Spain is consumed in the country, and the exportation is very insignificant compared with that of Cuba, Jamaica, or St Domingo.

Cotton, flax, and hemp, are not extensively raised,

and very little coffee is used in the country. Cocoa, vanilla, jalap, and tobacco, are cultivated; but of the latter there is a considerable importation from Havannah. Indigo is not produced in sufficient quantity for home consumption.

Since the middle of the sixteenth century, oxen, horses, sheep, and hogs, introduced by the conquerors, have multiplied surprisingly in all parts of New Spain, and more especially in the vast savannahs of the *provincias internas*. The exportation of hides is considerable, as is that of horses and mules.

Our common poultry have only of late years begun to thrive in Mexico; but there is a great variety of native gallinaceous birds in that country, such as the turkey, the hocco or curassow (*Crax nigra, C. globicera, C. pauxi*), penelopes, and pheasants. The Guinea fowl and common duck are also reared; but the goose is nowhere to be seen in the Spanish colonies.

The cultivation of the silkworm has never been extensively tried, although many parts of that continent seem favourable to it. An enormous quantity of wax is consumed in the festivals of the church; and, notwithstanding that a large proportion is collected in the country, much is imported from Havannah. Cochineal is obtained to a considerable amount.

Although pearls were formerly found in great abundance in various parts of America, the fisheries have now almost entirely ceased. The western coast of Mexico abounds in cachalots or spermaceti-whales (*Physter macrocephalus*); but the natives have hitherto left the pursuit of these animals to Europeans.

CHAPTER XXVI.

Mines of New Spain.

Mining Districts—Metalliferous Veins and Beds—Geological Relations of the Ores—Produce of the Mines—Recapitulation.

The mines of Mexico have of late years engaged the attention and excited the enterprise of the English in a more than ordinary degree. The subject is therefore one of much interest; but as later information may be obtained in several works, and especially in Ward's " Mexico in 1827," it is unnecessary to follow our author in all his details.

Long before the voyage of Columbus, the natives of Mexico were acquainted with the uses of several metals, and had made considerable proficiency in the various operations necessary for obtaining them in a pure state. Cortes, in the historical account of his expedition, states that gold, silver, copper, lead, and tin, were publicly sold in the great market of Tenochtitlan. In all the large towns of Anahuac gold and silver vessels were manufactured, and the foreigners, on their first advance to Tenochtitlan, could not refrain from admiring the ingenuity of the Mexican goldsmiths. The Aztec tribes extracted lead and tin from the veins of Tlacheo, and obtained cinnabar from the mines of Chilapan. From copper, found in the mountains of Zacotollan and Cohuixco, they manufactured their arms, axes, chisels, and other implements. With the use of iron they seem to have been unacquainted; but they contrived to give the requisite hardness to their tools by mixing a portion of tin with the copper of which they were composed.

At the period when Humboldt visited New Spain, it contained nearly 500 places celebrated for the metallic treasures in their vicinity, and comprehending nearly 3000 mines. These were divided into 37 districts, under the direction of an equal number of councils (*Diputaciones de mineria*), as follows:—

I. INTENDANCY OF GUANAXUATO.
1. Mining District of Guanaxuato.

II. INTENDANCY OF ZACATECAS.
2. Zacatecas.
3. Sombrerete.
4. Fresnillo.
5. Sierra de Pinos.

III. INTENDANCY OF SAN LUIS POTOSI.
6. Catorce.
7. Potosi.
8. Charcas.
9. Ojocaliente.
10. San Nicolas de Croix.

IV. INTENDANCY OF MEXICO.
11. Pachuca.
12. El Doctor.
13. Zuriapan.
14. Tasco.
15. Zacualpan.
16. Sultepec.
17. Temastaltepec.

V. INTENDANCY OF GUADALAXARA.
18. Bolanos.
19. Asientos de Ibarra.
20. Hostotipaquillo.

VI. INTENDANCY OF DURANGO.
21. Chihuahua.
22. Parral.
23. Guarisamey.
24. Cosiguiriachi.
25. Batopilas.

VII. INTENDANCY OF SONORA.
26. Alamos.
27. Copala.
28. Cosala.
29. San Francisco Xavier de la Huerta.
30. Guadalupe de la Puerta.
31. Santissima Trinidad de Pena Blanca.
32. San Francisco Xavier de Alisos.

VIII. INTENDANCY OF VALLADOLID.

33. Angangueo.
34. Inguaran.
35. Zitaquaro.
36. Tlalpajahua.

IX. INTENDANCY OF OAXACA.
37. Oaxaca.

X. INTENDANCY OF PUEBLA.
Several Mines.

XI. INTENDANCY OF VERA CRUZ.
Three Mines.

XII. OLD CALIFORNIA.
One Mine.

In the present state of the country the veins are the most productive, and the minerals disposed in beds or masses are very rare. The former are chiefly in primitive or transition rocks, rarely in secondary deposites. In the Old Continent granite, gneiss, and mica-slate, form the central ridges of the mountain-chains; but in the cordilleras of America these rocks seldom appear externally, being covered by masses of porphyry, greenstone, amygdaloid, basalt, and other trap-formations. The coast of Acapulco is composed of granite; and as we ascend towards the table-land of Mexico, we see it pierce the porphyry for the last time between Zumpango and Sopilote. Farther to the east, in the province of Oaxaca, granite and gneiss are visible in the high plains which are of great extent, traversed by veins of gold.

Tin has not yet been observed in the granites of Mexico. In the mines of Comarya syenite contains a seam of silver; while the vein of Guanaxuato, the richest in America, crosses a primitive clay-slate passing into talc-slate. The porphyries of Mexico are for the most part eminently rich in gold and silver. They are all characterized by the presence of hornblende and the absence of quartz. Common felspar is of rare occurrence, but the glassy variety is frequently observed in them. The rich gold mine of Villalpando, near Guanaxuato, traverses a porphyry, of which the basis is allied to clinkstone, and in which hornblende is extremely rare. The veins of Zuriapan intersect porphyries, having a greenstone basis, and contain a great variety of interesting minerals, such as fibrous zeolite, stilbite, grammatite, pycnite, native sulphur, fluor, barytes, corky asbestus, green garnets, carbonate and chromate of lead, orpiment, chrysoprase, and fire-opal.

Among the transition rocks, containing ores of silver, may be mentioned the limestone of the Real del Cardonal, Xacala, and Lomo del Toro, to the north of Zuriapan. In Mexico graywacke is also rich in metals.

The silver-mines of the Real de Catorce, as well as those of El Doctor and Xaschi, near Zuriapan, traverse Alpine limestone, which rests on a conglomerate with siliceous cement. In that and the Jura limestone are contained the celebrated silver-mines of Tasco and Tehuilotepec, in the intendancy of Mexico; and in these calcareous rocks the metalliferous veins display the greatest wealth.

It thus appears that the cordilleras of Mexico contain veins in a great variety of rocks, and that the deposites which furnish almost all the silver exported from Vera Cruz are primitive slate, graywacke, and Alpine limestone. The mines of Potosi in Buenos Ayres are contained in primitive clay-slate, and the richest of those of Peru in Alpine limestone. Our author here observes, that there is scarcely a variety of rock which has not in some country been found to contain metals, and that the richness of the veins is for the most part totally independent of the nature of the beds which they intersect.

Great advantage is derived in working the Mexican mines, from the circumstance that the most important of them are situated in temperate regions where the climate is favourable to agriculture. Guanaxuato is placed in a ravine, the bottom of which is somewhat lower than the level of the lakes of the valley of Mexico. Zacatecas and the Real de Catorce are a little higher; but the mildness of the air at these towns, which are surrounded by the richest mines in the world, is a contrast to the cold and disagreeable atmosphere of the Peruvian districts.

The produce of the Mexican mines is very unequally apportioned. The 2,500,000 marks, or 1,541,015 troy pounds of silver annually exported to Europe and Asia from Vera Cruz and Acapulco, are drawn from a very small number. Guanaxuato, Zacatecas, and Catorce, supply more than the half; and the vein of Guanaxuato alone yields more than a fourth part of the whole silver of Mexico, and a sixth of the produce of all America. The following is the order in which the richest mines of New Spain are placed, with reference to the quantity obtained from them:—

>Guanaxuato, in the intendancy of the same name.
>Catorce, in the intendancy of San Luis Potosi.
>Zacatecas, in the intendancy of the same name.
>Real del Monte, in the intendancy of Mexico.
>Bolanos, in the intendancy of Guadalaxara.
>Guarisamey, in the intendancy of Durango.
>Sombrerete, in the intendancy of Zacatecas.
>Tasco, in the intendancy of Mexico.
>Batopilas, in the intendancy of Durango.
>Zuriapan, in the intendancy of Mexico.
>Fresnillo, in the intendancy of Zacatecas.
>Ramos, in the intendancy of San Luis Potosi.
>Parral, in the intendancy of Durango.

The veins of Tasco, Sultepec, Tlapujahua, and Pachuca, were first wrought by the Spaniards. Those of Zacatecas were next commenced, and that of San Barnabe was begun in 1548. The principal one in Guanaxuato was discovered in 1558. As the total produce of all in Mexico, until the beginning of the eighteenth century, never exceeded 369,844 troy pounds of gold and silver yearly, it must be concluded, that during the sixteenth little energy was employed in drawing forth their stores.

The silver extracted in the thirty-seven districts was deposited in the provincial treasuries established in the chief places of the intendancies; and from the reports of these offices the quantity furnished by the different parts of the country may be determined. The following is an account of the receipts of eleven of these boards from the year 1785 to 1789:—

	Marks of Silver.
Guanaxuato,	2,469,000
San Luis Potosi,	1,515,000
Zacatecas,	1,205,000
Mexico,	1,055,000
Durango,	922,000
Rosario,	668,000
Guadalaxara,	509,000
Pachuca,	455,000
Bolanos,	364,000
Sombrerete,	320,000
Zuriapan,	248,000

Sum for five years, 9,730,000 = 5,997,633 troy pounds.

The mean produce of the mines of New Spain, including the northern part of New Biscay and those of Oaxaca, is estimated at above 1,541,015 troy pounds of silver,—a quantity equal to two-thirds of what is annually extracted from the whole globe, and ten times as much as is furnished by all the mines of Europe.

On the other hand the produce of the Mexican mines in gold is not much greater than those of Hungary and Transylvania; amounting in ordinary years only to 4315 troy pounds. In the former it is chiefly extracted from river-deposites by washing. Auriferous alluvia are common in the province of Sonora, and a great deal of gold has been collected among the sands with which the bottom of the valley of the Rio Hiaqui, to the east of the missions of Tarahumara, is covered. Farther to the north, in Pimeria Alta, masses of native gold weighing five or six pounds have been found. Part of it is also extracted from veins intersecting the primitive mountains. Veins of this metal are most frequent in the province of Oaxaca, in gneiss and mica-slate. The last rock is particularly rich in the mines of Rio San Antonio. Gold is also found pure, or mixed with silver-ore, in most of those which have been wrought in Mexico.

The silver supplied by the Mexican veins is extracted from a great variety of minerals. Most of it is obtained from sulphuretted silver, arsenical gray-copper, muriate of silver, prismatic black silver-ore, and red silver-ore. Pure or native silver is of comparatively rare occurrence.

Copper, tin, iron, lead, and mercury, are also procured in New Spain, but in very small quantities, although it would appear that they might be found to a great extent. The mercury occurs in various deposites, in beds, in secondary formations, and in veins traversing porphyries; but the amount obtained has never been sufficient for the process of amalgamation.

GOLD AND SILVER OF AMERICA.

The total value of gold and silver extracted from the mines of America, between 1499 and 1803, is estimated by Humboldt at 5,706,700,000 piasters, or (valuing the piaster at 4s. 4½d.) £1,248,340,625 sterling.

The annual produce of the mines of the New World, at the beginning of the present century, is estimated as follows:—

	Gold Marks.	Silver Marks.	Value in Dollars.
New Spain,	7,000	2,338,220	23,000,000
Peru,	3,400	611,090	6,240,000
Chili,	12,212	29,700	2,060,000
Buenos Ayres,	2,200	481,830	4,850,000
New Grenada,	20,505	. . .	2,990,000
Brazil,	29,900	. . .	4,360,000
	75,217	3,460,840	43,500,000

Valuing the dollar at 4s. 3d., the total annual produce would be £9,243,750.*

---

* According to Mr Ward (Mexico in 1827, vol. ii. p. 38), the annual average produce of the Mexican mines, before the revolution in 1810, amounted to 24,000,000 dollars, or £5,250,000, and the average exports to 22,000,000, or £4,812,500; but since the revolution the produce has been reduced to 11,000,000 dollars, or £2,406,250, while the exports in specie have averaged 13,587,052 dollars, or £2,970,198 each year. This reduction, it is unnecessary to say, has been caused by the unsettled state of the country, the emigration of the Old Spaniards, and the withdrawing of the funds which kept the mines in operation. In 1812, according to the same authority, the coinage had fallen to four and a half millions of dollars. It rose successively to six, nine, eleven, and twelve millions, which was the amount in 1819 in the capital alone. In 1820, the revolution in Spain caused a considerable fluctuation, and the coinage fell to 10,406,154 dollars. In 1821, when the separation from the mother-country became inevitable, the coinage sunk to five millions; from which it fell to three and a half, and continued in that state during 1823 and 1824. In 1825, the foreign capitals invested began to produce some effect; but in 1826, the total amount of coinage in the five mints of the Mexican republic did not exceed 7,463,300 dollars, or £1,632,594.

In 1827, seven English companies, one German, and two American, were employed in working mines in different parts of Mexico.

ENGLISH COMPANIES.

1. The Real del Monte Company, Captain Vetch, director, with an invested capital of £400,000.

To conclude our brief account of Humboldt's Political Essay on New Spain, it may be useful to present a few of the more interesting facts in the form of a recapitulation.

Physical Aspect.—Along the centre of the country runs a chain of mountains, having a direction from south-east to north-west, and afterwards from south to north. On the ridge or summit of this chain are extended vast table-lands or platforms, which gradually decline towards the temperate zone, their absolute height within the tropics being from 7545 to 7873 feet. The declivities of the cordilleras are wooded, while the central table-land is usually bare. In the equinoctial region the different climates rise as it were one above another from the shore, where the mean temperature is about 78°, to the central plains, where it is about 62°.

Population.—The whole population is estimated at 5,840,000, of which 4,500,000 are Indians, 1,000,000 creoles, and 70,000 European Spaniards.

---

2. The Bolanos Company, Captains Vetch and Lyon, directors, with a capital of £150,000.

3. Tlalpujahua Company, Mr De Rivafinola, director, with a capital of £180,000.

4. Anglo-Mexican Company, Mr Williamson, director; capital £800,000.

5. United Mexican Company; directors, Don Lucas Alaman, Mr Glennie, and Mr Agassis; capital £800,000.

6. The Mexican Company.

7. Catorce Company, Mr Stokes, director; invested capital not above £60,000.

At this period nearly three millions sterling of British capital were invested in the Mexican mines, or had been expended in enterprises immediately connected with them. The sudden change of feeling with respect to these adventures, which took place in England in 1826, had nearly put a stop to the operations commenced with so much energy; but confidence having been in some measure restored, it may be hoped that the mining companies will yet prove of great advantage both to Britain and to Mexico.

Agriculture.—The banana, manioc, maize, wheat, and potatoes, constitute the principal food of the people. The maguey or agave may be considered as the Indian vine. Sugar, cotton, vanilla, cocoa, indigo, tobacco, wax, and cochineal, are plentifully produced. Cattle are abundant on the great savannahs in the interior.

Mines.—The annual produce in gold is 4289 lb. troy; in silver, 1,439,832 lb.; in all, 23,000,000 of piasters (£5,031,250), or nearly half the quantity annually extracted from the mines of America. The mint of Mexico furnished from 1690 to 1803 more than 1,353,000,000 piasters (£295,968,750), and from the discovery of New Spain to the commencement of the nineteenth century, probably 2,028,000,000 piasters (£443,625,000). Three mining districts, Guanaxuato, Zacatecas, and Catorce, yield nearly half of all the gold and silver of New Spain.

Manufactures.—The value of the produce of the manufacturing industry of New Spain is estimated at 7,000,000 or 8,000,000 of piasters (valuing the piaster of exchange at 3s. $3\frac{1}{2}$d., £1,152,083 to £1,316,667). Cotton and woollen cloths, cigars, soda, soap, gunpowder, and leather, are the principal articles manufactured.

It is scarcely necessary to add, that the regions of America, which at the time of Humboldt's visit were Spanish colonies, have, after a series of sanguinary struggles, excited by the real or imagined grievances under which the inhabitants laboured, now succeeded in acquiring independence. This condition is more suitable than subjection to a remote power, protracted beyond the period at which such settlements are themselves fit to become empires. With colonies it is in some degree as with children. They receive the protection necessary for their growth, and obey at first from weakness and attachment; but beyond the stage at which they acquire a right to think for themselves, the attempt to perpetuate subordination necessarily excites a hatred which effectually quenches the feeble gratitude that man, in any condition, is capable of cherishing. The political divisions of America,—the land of republican principles,—are foreign to our object, and would require a more particular description than they could receive in this volume.

CHAPTER XXVII.

Passage from Vera Cruz to Cuba and Philadelphia, and Voyage to Europe.

Departure from Mexico—Passage to Havannah and Philadelphia—Return to Europe—Results of the Journeys in America.

LEAVING the capital of New Spain our travellers descended to the port of Vera Cruz, which is situated among sand-hills, in a burning and unhealthy climate. They happily escaped the yellow-fever,—which prevails there and attacks persons who have arrived from the elevated districts as readily as Europeans who have come by sea,—and embarked in a Spanish frigate for Havannah, where they had left part of their specimens. They remained there two months; after which they set sail for the United States, on their passage to which they encountered a violent storm that lasted seven days. Arriving at Philadelphia, and afterwards visiting Washington, they spent eight weeks in that interesting country, for the purpose of studying its political constitution and commercial relations. In August 1804 they returned to Europe, carrying with them the extensive collections which they had made during their perilous and fatiguing journeys.

The results of this expedition, conducted with so much courage and zeal, have been of the highest importance to science. With respect to natural history, it may be stated generally, that the mass of information already laid before the public, as obtained from the observation of six years, exceeds any thing that had been presented by the most successful cultivators of the same field during a whole lifetime. Much light has been thrown on the migrations and relations of the indigenous tribes of America, their origin, languages, and manners. The *Vues des Cordillières et Monumens des Peuples indigènes de l'Amérique*, 2 vols. folio, published in 1811, contains the fruit of researches into the antiquities of Mexico and Peru, together with the description of the more remarkable scenes of the Andes. It has been translated into English by Mrs H. M. Williams. The animals observed have been described in a work entitled *Recueil d'Observations de Zoologie et d'Anatomie Comparées, faites dans un Voyage aux Tropiques*, 2 vols. 4to.

In the department of botany the most important additions have been made to science. Our travellers brought with them to Europe an herbarium consisting of more than 6000 species of plants, and Bonpland's botanical journal contained descriptions of four thousand. The valuable works on this subject, that have appeared in consequence of the journey to America, form a new era in the history of botany. They are as follow:—

1. *Essai sur la Géographie des Plantes, ou Tableau Physique des Régions Equinoxiales, fondé sur des Observations et des Mesures faites depuis le 10me degrè de latitude australe, jusq'au 10me degrè de latitude boréale.* 4to.

2. *Plantes Equinoxiales Recueillies au Mexique,*

dans l'Ile de Cuba, dans les Provinces de Caracas, de Cumana, &c. 2 vols. fol.

3. *Monographie des Melastomes.* 2 vols. fol.

4. *Nova Genera et Species Plantarum.* 3 vols. fol.

5. *De Distributione Geographica Plantarum secundum Cœli Temperiem et Altitudinem Montium prolegomena.* 8vo.

The Essay on the Geography of Plants presents a general view of the vegetation, zoology, geological constitution, and other circumstances, of the equinoctial region of the New Continent, from the level of the sea to the highest summits of the Andes. The second work is by M. Bonpland, and contains methodical descriptions, in Latin and French, of the species observed; together with remarks on their medicinal properties and their uses in the arts. The Monography of the Melastomæ, which is also from the pen of M. Bonpland, contains upwards of 150 species of these plants, with others collected by M. Richard in the West Indies and French Guiana.

In his *Essai Géognostique sur le Gisement des Roches dans les deux Hémisphères,* published in 1826, and translated into English, Humboldt presents a table of all the formations known to geologists, and institutes a comparison between the rocks of the Old Continent and those of the cordillera of the Andes.

The astronomical treatises have been published in two quarto volumes, under the title of *Recueil d'Observations Astronomiques et de Mesures exécutées dans le Nouveau Continent.* This work contains the original observations made between the 12th degree of south latitude and the 41st degree of north latitude, transits of the sun and stars over the meridian, occultations of satellites, eclipses, &c.; a treatise on astronomical refractions under the torrid zone, considered as the effect of the decrement of caloric in the strata of the atmosphere; the barometric measurement of the Andes of Mexico, Venezuela, Quito, and New Grenada; together with a table of nearly 700 geographical positions. The greatest pains have been taken to verify the calculations. Our author presented to the *Bureau des Longitudes* his astronomical observations on the lunar distances and the eclipses of Jupiter's satellites, together with the barometrical elevations, which have been calculated and verified by M. Prony according to the formulæ of La Place.

In 1817 Humboldt laid before the *Académie des Sciences* his map of the Orinoco, exhibiting the junction of that river with the Amazon by means of the Casiquiare and Rio Negro.

The brief account of New Spain, which is presented in the preceding pages, has been extracted from the *Essai Politique sur la Nouvelle Espagne*, originally published in 2 vols. 4to, and translated into English. With respect to Humboldt's translators it may be remarked, that their want of scientific knowledge, and more especially of natural history, renders the English very much inferior to the French editions.

Most of the above-mentioned publications have appeared in the names of both travellers. The various works relating to the journey will make, when complete, twelve volumes in quarto, three in folio, two collections of geographical designs, and one of picturesque views. The detailed narrative of the expedition occupies four of these volumes; but an octavo edition has also been published under the title of *Voyage aux Régions Equinoxiales du Nouveau Continent, pendant les années* 1799, 1800, 1801, 1802, 1803, *et* 1804. The translation of this work by Mrs Williams is familiar to the English reader.

The labour necessary for reducing the observations made by our travellers to a condition fit for the public eye must have been very great; yet, possessed of a mind not less characterized by activity than the vastness of its acquirements, Humboldt in the mean while engaged in various investigations, which he has partly published in the foreign journals. In concert with M. Gay Lussac, with whom he lived for several years in the most intimate friendship, he has made numerous magnetic experiments, and verified Biot's theory respecting the position of the magnetic equator. They have found that the great mountain-chains, and even the active volcanoes, have no appreciable influence on the magnetic power; and have established the fact, that it gradually diminishes as we recede from the equator.

On the return of the philosophers from America, Bonpland was appointed by Bonaparte to the office of superintending the gardens at Malmaison, where the Empress Josephine, who was passionately fond of flowers, had formed a splendid collection of exotics. His amiable disposition, not less than his acquirements, procured for him the esteem of all who knew him. In 1818 he went to Buenos Ayres as Professor of Natural History. In 1820 he undertook an excursion to the interior of Paraguay; but when he arrived at St Anne on the eastern bank of the Parana, where he had established a colony of Indians, he was unexpectedly surrounded by a large body of soldiers, who destroyed the plantation and carried him off a prisoner. This was done by the orders of Dr Francia the ruler of Paraguay; and the only reason assigned was his having planted the tea-tree peculiar to that country, and which forms a valuable article of exportation. He was confined chiefly in Santa Martha, but was allowed to practise as a physician. Humboldt applied in vain for the liberation of his friend, for whom he appears to have cherished a sincere affection. According to a late report, however, he has obtained his liberty, and returned to Buenos Ayres.

In October 1818 our author was in London, where it was said that the allied powers had requested him to draw up a political view of the South American colonies. In November of the same year the King of Prussia granted him an annual pension of 12,000 dollars, with the view of facilitating the execution of a plan which he had formed of visiting Asia, and especially the mountains of Thibet. In the year 1822 he accompanied his majesty to the congress of Verona, and afterwards visited Venice, Rome, and Naples; and, in 1827 and 1828, delivered at Berlin a course of lectures on the physical constitution of the globe, which was attended by the royal family and the court. But, excepting the results of his investigations, which have appeared at intervals, we have no particular account of his occupations until 1829, when he undertook another important journey to the Uralian Mountains, the frontiers of China, and the Caspian Sea.

CHAPTER XXVIII.

Journey to Asia.

Brief Account of Humboldt's Journey to Asia, with a Sketch of the Four great Chains of Mountains which intersect the Central Part of that Continent.

No detailed narrative has yet been published of Humboldt's journey to Asiatic Russia; and the only sources of authentic information on the subject are to be found in a work lately printed at Paris, under the title of *Fragmens de Géologie et de Climatologie Asiatiques, par A. de Humboldt,* from which the following particulars are extracted:—

This illustrious traveller, accompanied by MM. Ehrenberg and Gustavus Rose, embarked at Nijnei-Novgorod on the Volga, and descended to Kasan and the Tartar ruins of Bolgari. From thence he went by Perm to Jekatherinenburg on the Asiatic side of the Uralian Mountains,—a vast chain composed of several ranges running nearly parallel to each other, of which the highest summits scarcely attain an elevation of 4593 or 4920 feet, but which, like the Andes, follows the direction of a meridian, from the tertiary deposites in the neighbourhood of Lake Aral to the greenstone rocks in the vicinity of the Frozen Sea. A month was occupied in visiting the central and northern parts of these mountains, which abound in alluvial beds containing gold and platina, the malachite mines of Goumeschevskoi, the great magnetic ridge of Blagodad, and the celebrated deposites at Mourzinsk, in which topaz and beryl are found. Near Nijnei-Tagilsk, a country which may be compared to Choco in South America, a mass of platina weighing about 21½ pounds troy has been found.

From Jekatherinenburg the travellers proceeded by Tioumen to Tobolsk on the Irtisch, and from thence by Tara, a steppe or desert of Baraba, which is dreaded on account of the torments caused by the multitudes of insects belonging to the family of *Tipulæ*, to Barnaoul on the banks of the Ob; the picturesque lake of Kolyvan; and the rich silver-mines of Schlangenberg, Riddersk, and Zyrianovski, situated on the south-western declivity of the Altaic range, the highest summit of which is scarcely so elevated as the Peak of Teneriffe. The mines of Kolyvan produce annually upwards of 49,842 troy pounds.

Proceeding southward from Riddersk to Oust-Kamenogorsk, they passed through Boukhtarminsk to the frontier of Chinese Zungaria. They even obtained permission to cross the frontier, in order to visit the Mongol post of Bates, or Khonimailakhou, northward of the Lake Dzaisang. Returning from this place to Oust-Kamenogorsk, they found the granite divided into nearly horizontal beds and overlying a slate-formation, the strata of which were partly inclined at an angle of 85° and partly vertical.

From Oust-Kamenogorsk they went along the steppe of the Middle Horde of the Kirghiz, by Semipolatinsk and Onisk and the lines of the Ichim Cossacks and Tobol, to reach the southern part of the Ural, where, in the vicinity of Miask, in a deposite of very small extent and at a depth of a few inches, were found three masses of native gold, two of which weighed 18·36 and the other 28·36 pounds troy.

They next proceeded along the Southern Ural to the fine quarries of green jasper at Orsk, where the river Jaik crosses the chain from east to west. From thence they passed by Souberlinsk to Orenburg, which notwithstanding its distance from the Caspian Sea is below the level of the ocean, and then visited the famous salt-mine of Iletzki, situated in the steppe of the Little Kirghiz Horde. They afterwards inspected the principal place of the Ouralsk Cossacks; the German colonies of the Saratov government on the left bank of the Volga; the great salt lake of Elton in the steppe of the Kalmucks; and a fine colony of Moravians at Sarepta; and, finally, arrived at Astracan. The principal objects of this excursion to the Caspian Sea were, the chemical analysis of its waters, which Mr Rose intended to make; the observation of the barometrical heights; and the collection of fishes for the great work of Baron Cuvier and M. Valenciennes.

From Astracan, the travellers returned to Moscow, by the isthmus which separates the Don and the Volga, near Tichinskaya, and the country of the Don Cossacks.

Of the heterogeneous materials composing the *Fragmens Asiatiques,* part only of which is from the pen of Humboldt, the memoir on the mountain-chains and volcanoes in the interior of Asia is the only one which can add any interest to our pages; the rest being of a character too strictly scientific. Of this paper a brief account is here given.

In our present state of knowledge volcanic phenomena are not to be considered as relating peculiarly to the science of geology, but rather as a department of general physics. When in action they appear to result from a permanent communication between the interior of the globe, which is in a state of fusion, and the atmosphere which envelopes the hardened and oxidated crust of our planet. Masses of lava issue like intermittent springs; and the superposition of their layers which takes place under our eyes bears a resemblance to that of the ancient crystalline rocks. On the crest of the cordilleras of the New World, as well as in the south of Europe and the western parts of Asia, an intimate connexion is manifested between the chemical action of volcanoes, properly so called, or those which produce rocks,—their form and position permitting the escape of earthy substances in a state of fusion,—and the mud-volcanoes of South America, Italy, and the Caspian Sea, which at one period eject fragments of rocks, flames, and acid vapours, and at another vomit muddy clay, naphtha, and irrespirable gases. There is even an obvious relation between the proper volcano and the formation of beds of gypsum and anhydrous rock-salt, containing petroleum, condensed hydrogen, sulphuret of iron, and, occasionally, large masses of galena; the origin of hot-springs; the arrangement of metallic deposites; earthquakes, which are ever and anon accompanied by chemical phenomena; and the sometimes sudden, and the sometimes very slow elevations of certain parts of the earth's surface.

This intimate connexion between these diversified appearances has of late years served to elucidate many problems in geology and physics which had previously been considered inexplicable. The analogies of observed facts, and the strict investigation of phenomena of recent occurrence, gradually lead us to more probable conjectures as to the events of those remote periods which preceded historical records. Volcanicity, or the influence which the interior of our planet exercises upon its external envelope in the various stages of its refrigeration, on account of the unequal aggregation in which its component substances occur, is, at the present day, in a very diminished condition; restricted to a small number of points; intermittent; simplified in its chemical effects; producing rocks only around small circular apertures, or over longitudinal cracks of small extent; and manifesting its power, at great distances, only dynamically, by shaking the crust of our planet in linear directions, or in spaces which remain the same during a great number of ages. Previous to the existence of the human race, the action of the interior of the globe upon the solid crust, which was increasing in volume, must have modified the temperature of the atmosphere, and rendered the whole surface capable of giving birth to those productions which ought to be considered as tropical, since, by the effect of the radiation and refrigeration of the exterior, the relations of the earth to a central body, the sun, began almost exclusively to determine the diversity of geographical latitudes.

In those primeval times, also, the elastic fluids, the volcanic powers of the interior, more energetic perhaps, and with more facility traversing the oxidated and solidified crust of the globe, filled this crust with crevices, and injected it with masses and veins of basalt, metallic substances, and other matters, introduced after the solidification of the planet had been completed. The period of the great geological revolutions was that when the communications between the fluid interior of the planet and its atmosphere were more frequent, acting upon a greater number of points, and when the tendency to establish these communications gave rise, in the line of the long crevices, to the cordilleras of the Andes and Himmaleh Mountains, the chains of less elevation, and the ridges whose undulations embellish the landscape of our plains. Our author then mentions, as proofs of these protrusions, the sandstone formations which extend from the plains of the Magdalena and Meta, almost without interruption, over platforms having an elevation varying from 8950 to 10,232 feet; and the bones of antediluvian animals intermingled on the summit of the Uralian chain of northern Asia with transported deposites, containing gold, diamonds, and platina. Another evidence of this subterranean action of elastic fluids, which heave up continents, domes, and mountain-chains, displace rocks and the organic remains which they contain, and produce eminences and depressions, is the great sinking of the ground which occurs in the west of Asia, of which the Caspian Sea and the Lake Aral form the lowest part (320 and 205 feet beneath the level of the ocean), but which extends far into the interior of the continent, stretching to Saratov and Orenburg on the Jaik, and probably to the south-east as far as the lower course of the Sihon (Jaxartes) and the Amou (the Oxus of the ancients). This depression of a continental mass extending to more than 320 feet below the surface of the ocean, he continues, has not hitherto obtained the necessary consideration which its importance demands, because it was not sufficiently known. It appears to him to have an intimate connexion with the upheaving of the Caucasian Mountains, those of Hindoo-kho, and of the elevated plain of Persia, which borders the Caspian Sea and the Mavar-ul-Nahar to the south; and, perhaps, more to the eastward, with the elevation of the great mass of land, which is designated by the vague and incorrect name of the central plain of Asia. This concavity he considers as a crater-country, similar to the Hipparchus, Archimedes, and Ptolemy, of the moon's surface, which have a diameter of more than 100 miles, and which may be rather compared with Bohemia than with our volcanic cones and craters.

In the course of the journey which Humboldt made in the summer of 1829 with MM. Ehrenberg and Rose, he passed in seven weeks over the frontiers of Chinese Zungaria, between the forts of Oust-Kamenogorsk, and Boukhtarminsk, and Khonimailakhou (a Chinese post to the north of the Lake Dzaisang), the Cossack line of the Kirghiz steppe, and the shores of the Caspian Sea. In the important commercial towns of Semipolatinsk, Petropalauska, Troitzkaia, Orenburg, and Astracan, he obtained from Tartars, Bucharians, and Tachkendis, information respecting the Asiatic regions in the vicinity of their native country. At Orenburg, where caravans of several thousand camels annually arrive, an enlightened individual, M. de Gens, has collected a mass of materials of the highest importance for the geography of Central Asia. Among the numerous descriptions of routes communicated by this person, our author found the following remark:—" In proceeding from Semipolatinsk to Jerkend, when we were arrived at the Lake Ala-koul or Ala-dinghiz, a little to the north-east of the great Lake Balkachi which receives the waters of the Ele, we saw a very high mountain which formerly vomited fire. Even now this mountain, which rises in the lake like a little island, occasions violent storms which incommode the caravans. For this reason some sheep are sacrificed to this old volcano by those who pass it."

This account, which was obtained from a Tartar who travelled at the commencement of the present century, excited a lively interest in our author, more especially as it brought to mind the burning volcanoes of the interior of Asia, made known through the researches of Abel Remusat and Klaproth in Chinese books, and whose great distance from the sea has excited so much surprise. Soon after his departure from Petersburg he received from M. de Klosterman, imperial director of police at Semipolatinsk, the following particulars, which were obtained from Bucharians and Tachkendis:—

"The route from Semipolatinsk to Kouldja is twenty-five days. It passes by the mountains Alachan and Rondegatay, in the steppe of the Middle Horde of the Kirghiz, the borders of the Lake Savande-koul, the Tarbagatai Mountains in Zungaria, and the river Emyl. When it has been traversed, the road unites with that which leads from Tchougeutchak to the province of Ele. From the banks of the Emyl to the Lake Ala-koul the distance is $39\frac{3}{4}$ miles. The Tartars estimate the distance of this lake from Semipolatinsk at 301 miles. It is to the right of the road, and extends from east to west $66\frac{1}{4}$ miles. In the midst of this lake rises a very high mountain, named Aral-toube. From this to the Chinese post, situated between the little Lake Janalache-koul and the river Baratara, on the banks of which reside Kalmucks, are reckoned 36 miles."

It is evident that the same mountain is alluded to in both these accounts; and with the view of connecting it with the volcanoes discovered by Klaproth and Abel Remusat mentioned in very ancient Chinese books, as existing in the interior of Asia to the north and south of Teen-shan, our author presents an account of the geography of this interesting region.

The middle and internal part of Asia, which forms neither an immense aggregate of hills nor a continuous platform, is intersected from east to west by four great systems of mountains, which have exercised a decided influence upon the movements of nations. These systems are: 1. The Altaic, which is terminated to the west by the mountains of the Kirghiz; 2. Teen-shan; 3. Kwan-lun; and, 4. The Himmaleh chain. Between the Altaic range and Teen-shan are Zungaria and the basin of the Ele; between Teen-shan and Kwan-lun, Little or Upper Bucharia, or Cashgar, Yarkand, Khoten, or Yu-thian, the great desert, Toorfan, Khamil, and Tangout, or the Northern Tangout of the Chinese, which must not be confounded with Thibet or Sefan. Lastly, between Kwan-lun and the Himmaleh are Eastern and Western Thibet, in which are Lassa and Ladak. Were the three elevated plains situated between the Altai, Teen-shan, Kwan-lun, and the Himmaleh, to be indicated by the position of three Alpine lakes, we might select for this purpose those of Balkachi, Lop, and Tengri, which correspond to the plains of Zungaria, Tangout, and Thibet.

1. *System of the Altai.*—It surrounds the sources of the Irtisch and Jenisei or Rem. To the east it takes the name of Tangnou; between the lakes Rossogol and Baikal, that of the Sayanian Mountains; beyond this it takes the name of Upper Kentai, and the Davourian Mountains; and, lastly, to the north-east it connects itself with the

Jablonnoikhrebet chain, Khingkhan, and the Aldan Mountains, which advance along the Sea of Ochotzk. The mean latitude of its prolongation from east to west is between 50° and 51° 30′. The Altaic range, properly so called, scarcely occupies seven degrees of longitude; but the northern part of the mountains, surrounding the great mass of elevated land in the interior of Asia, and occupying the space comprised between 48° and 51°, is considered as belonging to this system, because simple names are more easily retained by the memory, and because that of Altai is more known to Europeans by its great metallic richness, which amounts annually to 45,907 troy pounds of silver, and 1246 troy pounds of gold. The Altaic Mountains are not a chain forming the boundary of a country like the Himmaleh, which limit the elevated plain of Thibet, and have a rapid slope only on the side next to India, which is lower. The plains in the neighbourhood of the Lake Balkachi have not an elevation of more than 1920 feet above the sea.

Between the meridians of Oust-Kamenogorsk and Semipolatinsk the Altaic system is prolonged, from east to west under the parallels of 49 and 50 degrees by a chain of low mountains, over an extent of 736 miles, as far as the steppe of the Kirghiz. This ridge has been elevated through a fissure which forms the line of separation of the streams of the Sara-sou and Irtisch, and which regularly follows the same direction over an extent of 16 degrees of longitude. It consists of stratified granites not intermixed with gneiss, and of greenstone, porphyry, jasper, and transition-limestone, in which there occur various metallic substances. This low range does not reach the southern extremity of the Ural, a chain which, like the Andes, presents a long wall running north and south, with metallic mines on its eastern slope, but terminates abruptly in the meridian of Sverinogovloskoi.

Here commences a remarkable region of lakes, comprising the group of Balek-koul (lat. 51° 30′), and that of Koumkoul (lat. 49° 45′), indicating an ancient communication of a mass of water with the Lake Ak-sakal, which receives the Tourgai and the Kamichloi Irghiz, as well as with the Lake Aral; and which would seem from Chinese accounts to have formed part of a great plain extending to the borders of the Frozen Sea.

2. *System of Teen-shan.*—The mean latitude of this system is the 42d degree. Its highest summit is perhaps the mass of mountains covered with perpetual snow, and celebrated under the name of Bokhda-ovla, from which Pallas gave the designation of Bogdo to the whole chain. From Bokhda-ovla and Khatoun-bokhda, the Teen-shan mountains run eastward towards Bar-koul, where they are suddenly lowered so as to fall to the level of the elevated desert, called the Great Gobi or Cha-mo, which extends from Koua-tcheou, a Chinese town, to the sources of the Argoun. If we now return to Bokhda-ovla, we find the western prolongation of these mountains stretching to Goudja and Koutche, then between Lake Temoustou and Aksou to the north of Cashgar, and running towards Samarcand. The country comprehended between the Altaic chain and the Teen-shan mountains is shut up to the east, beyond the meridian of Pekin, by the Khingkhan-ovla, a lofty ridge, which runs from south-west to north-east; but to the west it is entirely open.

2 c

The case is very different with the country limited by the second and third systems, the Teen-shan and Kwan-lun ranges; it being closed to the west by a transverse ridge, which runs north and south, under the name of Bolor or Belour-tagh. This chain separates Little Bucharia from Great Bucharia, the country of Cashgar, Badakshan, and Upper Djihoun. Its southern part, which is connected with the Kwan-lun system, forms a part of the Tsungling of the Chinese. To the north it joins the chain which passes to the north-west of Cashgar. Between Khokand, Dervagel, and Hissar, consequently between the still unknown sources of the Sihon and Amou-deria, the Teen-shan rises before lowering again in the Kanat of Bochara, and presents a group of high mountains, several of which are covered with snow even in summer. More to the east it is less elevated. The road from Semipolatinsk to Cashgar passes to the east of Lake Balkachi and to the west of Lake Ossi-koul, and crosses the Narim, a tributary of the Sihon. At the distance of $69\frac{1}{2}$ miles from the Narim to the south, it passes over the Rovat, which has a large cave, and is the highest point before arriving at the Chinese post to the south of the Ak-sou, the village of Artuche, and Cashgar. This city, which is built on the banks of the Ara-tumen, has 15,000 houses and 80,000 inhabitants, although it is smaller than Samarcand.

The western prolongation of the Teen-shan or the Mouz-tagh, is deserving of particular examination. At the point where the Bolor or Belour-tagh joins the Mouz-tagh at right angles, the latter continues to run without interruption from east to west, under the name of Asferah-tagh, to the south of the Sihon, towards Kodjend and Ourat-eppeh in Ferganah. This chain of Asferah, which is covered with perpetual snow, separates the sources of the Sihon (Jaxartes) from those of the Amou (Oxus). It turns to the south-west nearly in the meridian of Kodjend, and in this direction is named, till it approaches Samarcand, Aktagh, or Al-Botous. More to the west, on the fertile banks of the Kohik, commences the vast depression of ground comprising Great Bucharia and the country of Mavar-ul-Nahar; but beyond the Caspian Sea, nearly in the same latitude and in the same direction as the Teen-shan range, is seen the Caucasus with its porphyries and trachytes. It may, therefore, be considered as a continuation of the fissure upon which the Teen-shan is raised in the east, just as, to the west of the great mass of mountains of Adzarbaidjan and Armenia, Mount Taurus is a continuation of the action of the fissure of the Himmaleh and Hindoo-Coosh mountains.

3. *Kwan-lun System.*—The Kwan-lun or Koul-koun chain is between Khoten, the mountains of Khoukhou-noor and Eastern Thibet, and the country named Katchi. It commences to the west at the Tsung-ling mountains. It is connected with the transverse chain of Bolor, as observed above, and, according to the Chinese books, forms its southern part. This corner of the globe, between Little Thibet and the Boda Kohan, is very little known, although it is rich in rubies, lapis lazuli, and mineral turquois; and, according to recent accounts, the plain of Khorassan, which runs in the direction of Herat, and limits the Hindoo-kho to the north,

appears to be rather a continuation of the Tsung-ling and of the whole system of Kwan-lun to the west, than a prolongation of the Himmalehs, as is commonly supposed. From the Tsung-ling the Kwan-lun, or Koulkoun range, runs from west to east towards the sources of the Hoang-ho or Yellow River, and penetrates with its snowy summits into Chen-si, a province of China. Nearly in the meridian of these springs rises the great mass of mountains on the Lake Khoukhou-noor, resting to the north upon the snowy chain of the Nanshan or Ki-leen-shan, which also runs from west to east. Between Nanshan and Teen-shan, the heights of Tangout limit the margin of the upper desert of Gobi, or Cha-mo, which is prolonged from south-west to north-east. The latitude of the central part of the Kwan-lun range is 35° 30'.

4. *Himmaleh System.*—This system separates the valleys of Cashmere and Nepaul from Bootan and Thibet. To the west it rises in the mountain Javaher to an elevation of 25,746 feet, and to the east in Dhwalagiri to 28,074 feet above the level of the sea. Its general direction is from north-west to south-east, and thus it is not at all parallel to the Kwan-lun range, to which it approaches so near in the meridian of Attok and Jellalabad that they seem to form the same mass of mountains. Following the Himmaleh range eastward, we find it bordering Assam on the north, containing the sources of the Brahmapoutra, passing through the northern part of Ava, and penetrating into Yun-nan, a province of China, to the west of Young-tchang. It there exhibits pointed and snow-clad summits. It bends abruptly to the north-east, on the confines of

Hou-kouang, Kiang-si, and Foukian, and advances its snowy peaks towards the ocean; the island of Formosa, the mountains of which are in like manner covered during the greater part of summer, being its termination. Thus we may follow the Himmaleh system as a continuous chain from the Eastern Ocean, through Hindoo-kho, across Candahar and Khorassan, to beyond the Caspian Sea in Adzarbaidjan, along an extent of 73 degrees, or half the length of the Andes. The western extremity, which is volcanic (like the eastern part), loses its character of a chain in the mountains of Armenia, which are connected with Sangalou, Bingheul, and Kachmirdaugh, in the pashalic of Erzeroum. The mean direction of the system is north 55° west.

These mountain-chains, with their various ramifications and intervening platforms and valleys, afford evidence to our author of revolutions anciently undergone by the crust of the globe; these having been elevated by matter thrust up in the line of enormous cracks and fissures. The great depression of Central Asia, spoken of above, he considers as having been caused by the same action. Analogous to the Caspian Sea and other cavities in this district, are the lakes formed in Europe at the foot of the Alps, and which also owe their origin to a sinking of the ground. It is chiefly in the extent of this depression of Central Asia, and consequently in the space where the resistance was least, that we find traces of volcanic action. Several volcanoes are described in this space by ancient Chinese writers, who also mention a variety of volcanic products, such as sal ammoniac and sulphur, which form articles of commerce.

" We thus know," says our author, " in the interior of Asia, a volcanic territory, the surface of which is upwards of 2500 square geographical miles, and which is from 1000 to 1400 miles distant from the sea. It fills the half of the longitudinal valley situated between the first and second system of mountains. The principal seat of volcanic action appears to be in the Teen-shan. Perhaps the colossal Bokhda-ovla is a trachytic formation like Chimborazo." On both sides of the Teen-shan violent earthquakes occur. The city of Aksou was entirely destroyed at the commencement of the eighteenth century by a commotion of this nature. In Eastern Siberia the centre of the circle of shocks appears to be at Irkutzk, and in the deep basin of the Baikal Lake, in the vicinity of which volcanic products are observed. But this point of the Altaic range is the extreme limit of these phenomena, no earthquakes having been experienced farther to the west, in the plains of Siberia, between the Altaic and Uralian ranges, or in any part of the latter.

The volcanic territory of Bichbalik is to the east of the great depression of Asia. To the south and west of this internal basin we find two cones in activity,—Demavend, which is visible from Teheran, and Seiban of Ararat, which is covered with vitreous lavas. On both sides of the isthmus between the Caspian and the Black Sea springs of naphtha and mud-eruptions are numerous.

On the western margin of the great depression, if we proceed from the Caucasian isthmus to the north and north-west, we arrive at the territory of the great horizontal and tertiary deposites of Southern Russia and Poland. Here we find igneous rocks piercing the red sandstone of Jekaterinoslav, together with asphaltum and springs impregnated with sulphurous gases.

A phenomenon so great as that of the central depression of Asia, which resembles the circular valleys of the moon, could have been produced only by a very powerful cause acting in the interior of the earth. This cause, while forming the crust of the globe by sudden raisings and sinkings, probably filled with metallic substances the fissures of the Uralian and Altaic chains.

It is not the custom of our author to detail personal adventures, his object being to give a scientific character to his narrative; and for this reason his relations may be less interesting to many readers than some of the travels and voyages which have of late been so profusely offered to the public. He is at present engaged in preparing an account of his Asiatic tour, the full details of which will appear under the general title of " A Journey to the Uralian Range, the Mountains of Kolyvan, the Frontier of Chinese Zungaria, and the Caspian Sea, made by Order of the Emperor of Russia, in 1829, by A. de Humboldt, G. Ehrenberg, and G. Rose." It will consist of three distinct works: 1. A geological and physical view of the north-west of Asia, observations of terrestrial magnetism, and results of astronomical geography, by Baron Humboldt. 2. The mineralogical and geological details, the results of chemical analysis, and the narrative of the journey, by M. Rose. 3. The botanical and zoological part, with observations on the distribution of plants and animals, by M. Ehrenberg.

Any formal eulogy on our illustrious author must be altogether unnecessary, for his renown has extended over all parts of the civilized world, and, at the present day, there is not a man of science in Europe whose name is more familiar. Long after his career shall have terminated, he will be remembered as one of the chief ornaments of an age peculiarly remarkable in the history of the world. As there is a natural desire in most people to become acquainted with the physical tenement of a mind whose productions have excited interest, or afforded useful knowledge, the publishers have endeavoured to gratify it in some measure, by prefixing a portrait of this distinguished philosopher in his younger days. It were easy to point out in this delineation the most decided marks of that capacious intellect and gentleness of disposition,—that combination of power and benignity,—by which he is characterized; but the physiognomist needs no assistance in a matter of this kind, for when the character is known, it is easy to read it in the features.

THE END.

For EU product safety concerns, contact us at Calle de José Abascal, 56–1°,
28003 Madrid, Spain or eugpsr@cambridge.org.

 www.ingramcontent.com/pod-product-compliance
Ingram Content Group UK Ltd.
Pitfield, Milton Keynes, MK11 3LW, UK
UKHW010352140625
459647UK00010B/1016